D1238542

Icy Worlds of the Solar System

Earth is the only planet known to have liquid water, and water ice has been present over parts of the Earth for much of its history. Scientists have only recently come to understand how widespread the presence of ice is in our solar system. Deposits of water ice may exist in unexpected places, such as in the polar craters of Mercury, the closest planet to the Sun. Other ices, such as methane ice and nitrogen ice, abound in our solar system. These ices play an important role in the geological and atmospheric characteristics of the bodies in our solar system.

This book focuses on the occurrence and significance of water ice, and ices formed by other materials, in the solar system. The findings discussed are the result of three decades of spacecraft exploration of the planets, complemented by ground- and space-based observations. It considers the implications of the reservoirs of water ice for the presence of life elsewhere in our solar system, and for habitability by human explorers who may venture to these distant worlds in the future. Written at an accessible level, this book will be of interest to students and professionals in planetary science, geology, and related areas.

PAT DASCH is a consultant in the space industry specializing in policy and public outreach. She has published and broadcast on a wide variety of space-related topics for the past 20 years. Dasch has held a variety of positions, including Executive Director of the Washington DC-based National Space Society, and Editor in Chief of *Ad Astra*, the bi-monthly magazine of the National Space Society. She worked in solar system exploration at NASA Headquarters and has also acted as Director of the Dial-A-Shuttle broadcast program at the Johnson Space Center, and served as a research associate at the Lunar and Planetary Institute.

Icy Worlds of the Solar System

Edited by

Pat Dasch

CAMBRIDGE
UNIVERSITY PRESS

PUBLISHED BY THE PRESS SYNDICATE OF THE UNIVERSITY OF CAMBRIDGE
The Pitt Building, Trumpington Street, Cambridge, United Kingdom

CAMBRIDGE UNIVERSITY PRESS
The Edinburgh Building, Cambridge, CB2 2RU, UK
40 West 20th Street, New York, NY 10011–4211, USA
477 Williamstown Road, Port Melbourne, VIC 3207, Australia
Ruiz de Alarcón 13, 28014 Madrid, Spain
Dock House, The Waterfront, Cape Town 8001, South Africa

http://www.cambridge.org

First published 2004

Printed in the United Kingdom at the University Press, Cambridge

Typeface Swift Regular 9.75/14 pt. *System* LaTeX 2_ε [TB]

A catalog record for this book is available from the British Library

Library of Congress Cataloging in Publication data
Icy worlds of the Solar System / edited by Pat Dasch
p. cm.
Includes bibliographical references and index.
ISBN 0 521 64048 2
1. Solar system. 2. Ice. I. Dasch, Pat.
QB505.I39 2004
523.2 – dc22 2003069573
ISBN 0 521 64048 2 hardback

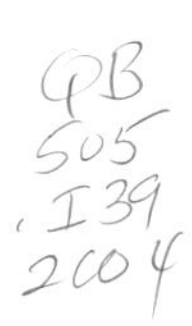

Contents

Contributors

ROBERT BINDSCHADLER

Dr. Robert Bindschadler, a glaciologist at the NASA Goddard Space Flight Center, has been an active Antarctic field researcher for the past 20 years, has led 11 field expeditions to Antarctica, and has participated in many others to glaciers and ice caps around the world. He maintains an active interest in the dynamics of glaciers and ice sheets, primarily on Earth, investigating how remote sensing can be used to improve our understanding of the role of ice in the Earth's climate. He has advised the US Congress and the Vice President on the stability of ice sheets and ice shelves and served on many scientific commissions and study groups as an expert in glaciology and remote sensing of ice. Some of the more significant awards he has received are: Excellence in Federal Career (1989), the Antarctic Service Medal (1984), and the NASA Exceptional Scientific Achievement Medal (1994), and he was appointed a Goddard Senior Fellow (2000). He has published over 100 scientific papers including numerous review articles and has appeared on television and radio to comment on glaciological impacts of the climate on the world's ice sheets and glaciers. He currently is President of the International Glaciological Society, chairs the West Antarctic Ice Sheet Initiative, sits on the Editorial Board of the American Geophysical Union's Antarctic Research Series, and is an editor for the *Journal of Glaciology*.

BRYAN BUTLER

Bryan Butler was born and raised in northern Utah, where he attended undergraduate school at Utah State University. He graduated in 1988, earning BS degrees (Magna Cum Laude) in Electrical Engineering and Computer Science. From there, he went on to graduate work at Caltech, earning his MS in 1991 and his Ph.D. in 1994, both in Planetary Science. Since then he has been an employee of the National Radio Astronomy Observatory as a member of the scientific staff. His research focuses on the radio emission and radar scattering from solar system bodies, including the planets, moons, asteroids, and comets, and the theory and implementation of radio interferometry.

DALE CRUIKSHANK

Dale Cruikshank is a Research Scientist in the Astrophysics Branch at NASA Ames Research Center. Prior to taking this position with NASA in 1988, he had been a tenured research faculty member at the University of Hawaii for nearly 18 years. Cruikshank's principal work is in the study of the surfaces and atmospheres of the planets and small bodies of the solar system, using the techniques of infrared spectroscopy, photometry, and radiometry, in which he has authored or coauthored about 250 research papers.

Cruikshank and his colleagues have discovered many of the ice species found on Pluto, Triton, and the satellites of Saturn and Uranus. Seeking to understand the dark red materials covering many satellites, comets, Kuiper Belt objects, Centaurs, and asteroids, Cruikshank has advanced the idea, with spectroscopic observational support, that complex organic material is an important constituent of the surface materials of these bodies. Cruikshank is a scientist on the Cassini mission to Saturn, the Space Infrared Telescope Facility (SIRTF), and the New Horizons mission to Pluto-Charon and the Kuiper Belt.

JONATHAN LUNINE

Jonathan I. Lunine is Professor of Planetary Sciences and of Physics, and chairs the Theoretical Astrophysics Program at the University of Arizona. His research interests center broadly on the formation and evolution of planets and planetary systems. He is an interdisciplinary scientist on the Cassini mission to Saturn, and on the James Webb Space Telescope. Dr. Lunine is the author of the book *Earth: Evolution of a Habitable World* (Cambridge University Press, 1999). He is a fellow of the American Association for the Advancement of Science and of the American Geophysical Union, which awarded him the James B. Macelwane medal, and winner of the Harold C. Urey Prize (American Astronomical Society) and Ya. Zeldovich Award of COSPAR's Commission B. Dr. Lunine earned a BS in Physics and Astronomy from the University of Rochester in 1980, followed by MS (1983) and Ph.D. (1985) degrees in Planetary Science from the California Institute of Technology.

MICHAEL MELLON

Michael T. Mellon is a planetary scientist at the Laboratory for Atmospheric and Space Physics at the University of Colorado at Boulder. His research has concentrated on the distribution and behavior of ice on Mars and he has made a detailed study of the similarities and differences between martian and terrestrial permafrost.

TOBY OWEN

Tobias Owen is a Professor of Astronomy at the Institute for Astronomy of the University of Hawaii. He is studying planets, satellites, and comets with the telescopes on Mauna Kea and by means of deep space exploration. Having participated in the Viking, Voyager, and Galileo missions, Owen is presently an Interdisciplinary Scientist and member of the Probe Mass Spectrometer, Aerosol Collection, and Pyrolysis and Consolidated Infrared Spectrometer Teams on the NASA–ESA Cassini Huygens mission to the Saturn system. He is also participating in the ESA Rosetta mission to Comet Churyumov-Gerasimenko.

PAUL SCHENK

Dr. Paul Schenk is a planetary geologist at the Lunar and Planetary Institute in Houston, Texas. He graduated from Washington University in Saint Louis and has been at the LPI since 1991, where he studies the geology and cratering history of the icy satellites of the outer solar system. Chief among his interests is mapping the topography of these bodies (and other planets) using stereo imaging and shape-from-shading techniques. He is currently completing topographic mapping of the large jovian satellites using available Voyager and Galileo imaging data, and preliminary topographic mapping of Saturn's satellites in preparation for Cassini's tour of the Saturn system beginning in late 2004.

JOHN STANSBERRY

John Stansberry is an assistant astronomer at Steward Observatory, the University of Arizona, in Tucson. He is a member of the Multiband Imaging Photometer for SIRTF (MIPS) instrument team. MIPS is the most sensitive camera ever built for use in the far-IR 20–200 micron range, and was launched as a part of the Space Infrared Telescope Facility (SIRTF) on August 25, 2003. MIPS will be capable of accurately measuring the heat given off, not only by Pluto and Triton, but also by KBOs as small as 200 kilometers in diameter, and will also be useful for many astronomical observations.

Dr. Stansberry obtained his BA in Physics from Colorado College in 1985, and his Ph.D. in Planetary Science from the University of Arizona in 1995. His dissertation was on the interactions of the surfaces and atmospheres of Triton and Pluto. After his doctoral work, he has participated in an extensive program to observe Jupiter's moon, Io, in the infrared, mapping the locations, intensity, and variability of volcanic eruptions. He also was a member of the runner-up proposal to send a mission to Pluto and the Kuiper Belt, leading the development of a thermal imager that would have mapped surface temperatures on Pluto and Kuiper Belt objects.

PAT DASCH

Pat Dasch is a consultant to the space industry on space policy and public outreach issues and a writer who publishes on a wide variety of space-related topics. Recent projects include development of content for a PBS program on the future of human spaceflight, and development of strategic positions related to the future of internationally coordinated space missions for the Space Policy Summit held in Houston, TX, in October 2002.

She is editor in chief of the 4-volume reference work *Space Science for Students*, published in September 2002 by Macmillan Reference. Ms. Dasch has authored numerous articles on space exploration, and presented testimony to Congress. *Images of Earth*, co-authored with Peter Francis for George Philip (UK) and Prentice Hall in the US, in 1984, won the Geographical Society of Chicago prize for best remote sensing publication in 1985 and was carried by the Aviation Week Book Club. Ms. Dasch has also produced a number of educational slide sets, the most recent being "Life on Mars???" and "Asteroids."

Previously, Ms. Dasch was Executive Director of the National Space Society (1997–2001). From 1994–98 she was Editor in Chief of *Ad Astra* magazine, the magazine of the National Space Society, and from 1988–94 worked for SAIC as a planetary science analyst in the Solar

System Exploration Division at NASA Headquarters, Washington, DC. From 1983–87 she was director of the Dial-A-Shuttle program at the Johnson Space Center and a research associate at the Lunar and Planetary Institute in Houston working on shuttle Earth observations imagery.

Preface

In the last decade, information from the Galileo mission to the Jupiter system and advances in ground-based astronomy have greatly enhanced our reservoir of knowledge about ices in our solar system. An acceleration in the search for signs of life in the solar system (and the water that is necessary for the Earth-based life forms that we are familiar with) that followed the 1996 discovery of possible meteoritic evidence for the existence of ancient life on Mars, together with the technological revolution in both space- and Earth-based sensors, has resulted in significant new developments in understanding of the pervasive presence and geological significance of ices in our solar system.

In this book, recognized planetary experts interpret the role and impact of ice in our corner of the universe and debate the many outstanding questions that remain to be answered. Each chapter contains exciting, cutting-edge information revealing the complexity and wonder of the universe in which we live.

The findings from the Opportunity and Spirit rovers (Mars), the SMART-1 mission on route to the Moon and Cassini (saturnian system) will help to answer some of the outstanding questions about ice and will undoubtedly reveal new conundrums for our future contemplation.

Other missions planned for the future, such as the MESSENGER and BepiColombo missions to Mercury, and the New Horizons mission to Pluto and the Kuiper Belt, will continue this second wave of planetary exploration with rigorous surveys and collection of detailed scientific data.

Acknowledgements

First and foremost I want to thank the contributors to this book – Robert Bindschadler, Bryan Butler, Dale Cruikshank, Jonathan Lunine, Michael Mellon, Toby Owen, Paul Schenk, and John Stansberry – whose state-of-the-science knowledge, insights, and curiosity made this book possible.

My deepest gratitude to my husband, Julius, for his sustained support, encouragement, and interest throughout the evolution of this book, which appeared to be in constant revision right up to manuscript submission, as the scientific story continually rewrote itself with a steady stream of new and fascinating scientific developments.

My thanks to Simon Mitton, Jacqueline Garget, Carol Miller and Joseph Bottrill at Cambridge University Press for their perseverance and professional management of this project.

JONATHAN I. LUNINE
University of Arizona

Introduction

We live in unusual times for the Earth: over the last 0.1% of our planet's history, extensive caps of ice have covered the poles and vast glaciers have advanced and retreated across the continents. The effects on the Earth have been significant: the landscape and climate patterns have been reshaped, leading to extinctions and the birth of new species (including, perhaps, our own). For much of Earth's history before that, the planet was nearly ice free. Whether our own technological activities will drive the Earth back toward that state remains a politically charged and unanswered question.

While water ice is an ephemeral feature over much of the Earth for much of its history, it is an important and permanent part of most of the other objects in our solar system. This book treats the occurrence and significance of water ice and ices formed by other materials in the solar system. The findings discussed in the chapters are the results of almost four decades of spacecraft exploration of the planets, complemented by ground-based observations which themselves have been revolutionized by progress in electronic detectors, large telescopes, and airborne platforms.

Water ice exists at the Martian poles and just beneath the surface at high latitudes across the planet. It is known to be present on the surfaces of many of the moons of the giant planets, on Pluto and in comets, and less direct information suggests it to be a key component of the interiors of these bodies. The surfaces of the distant Kuiper Belt objects, thought to be leftovers from planet formation, probably have water ice as well, though a definitive answer will require continued observations using the world's largest telescopes. Water has even been detected, from the European Infrared Space Observatory, in the frigid atmosphere of Saturn's moon Titan. In addition, radar studies infer ice to be present at the poles of searing Mercury, protected in the shadowed floors of craters. The same may be true for Earth's Moon, where the Lunar Prospector spacecraft has indirectly detected the presence of water at the poles.

Why is water ice so ubiquitous in our solar system? The answer lies in the nature of the material from which the planets formed, the conditions prevailing in the solar system during and after formation, and in the properties of water itself. The elements from which the solar system (including ourselves) was made, other than the most abundant hydrogen and helium, originated in the nuclear fusion furnaces of a previous generation of stars. Oxygen is among the most common products of hydrogen fusion in many stars, and hence is ubiquitous in the molecular clouds from which new stars are born. Oxygen readily combines with hydrogen to make H_2O, or water. Other atoms with which oxygen can combine such as silicon, to make rocky minerals, are much less abundant than hydrogen; hence water is the most plentiful oxygen-bearing species in the cosmos.

To go from water molecules in the gas phase to water ice requires the right conditions. Astronomers observe disks of gas and dust around some newly formed or forming stars, and it is believed that such disks are the progenitors of planets. Simple models of disks show quite generally that they must be hot close to the central star and cooler farther out. Models of our own solar system in its primordial, or "solar nebula" phase, indicate that the gas beyond what is now the asteroid belt was cool enough to allow water to condense out in the form of water ice grains. These grains agglomerated into larger objects and became the seeds of giant planets, their moons and other bodies of the outer solar system. In the asteroid belt and inward in the disk, the gas was simply too hot to allow water to condense out but silicate (rocky) and metallic grains were present. These were the building blocks of the terrestrial planets Mercury, Venus, Earth, Mars, and Earth's Moon.

Other molecules in the gas of our progenitor disk could not condense out, or did so only at very great distances from the Sun (and hence extremely low temperatures) because of their particular chemical properties. These molecules, methane, carbon monoxide, nitrogen, differ greatly from water in terms of their tendency to form liquids and solids only at very low temperatures. Water condenses out at moderate temperatures because of a peculiar quality of the molecule, called hydrogen bonding, which also turns out to be important in water's critical function as a transport medium in biological processes. Without this property, the solar system from Jupiter outward might well have been much emptier than it in fact is, with only rocky grains available to form solid bodies. Even the giant planets themselves might have been absent or smaller, since the high abundance of ice grains in the solar nebula is thought to have quickly initiated their growth in the limited time available to assemble the planets.

Water ice in the outer solar system trapped many other kinds of molecules in naturally occurring pore spaces. As Owen describes in Chapter 3, icy bodies such as comets may have delivered water and other biologically important molecules such as methane to the Earth early in its history. The vast amount of water Earth received set our planet on a trajectory unique among all the solar system's planets – a world habitable for billions of years such that life could gain a foothold, evolve, and diversify. In Chapter 1, Bindschadler describes the myriad forms of water on Earth and its profound effects on our home world.

Moving from the Earth to the inner planets, water ice cannot exist today on the torrid surface of Venus, though our "sister" planet might well have played host to liquid water very early in its history. We expect no water on sunbaked Mercury either, but Butler discusses the remarkable discovery that the poles of Mercury are highly radar reflective and hence might contain water ice (Chapter 2). Because Mercury's poles are always perpendicular to the planet's orbit, crater floors right at or near the poles are cold enough to retain water. It is then natural to ask whether our own Moon might play host to water ice at its poles, and Butler discusses the evidence, pro and con, from the two recent robotic missions Clementine and Prospector.

Mars certainly received its share of cometary water ice, and the early climate of the red planet may have been warm enough to allow liquid water, and perhaps even life, to exist for a time. In Chapter 4, Mellon reviews the search for evidence of this ancient warm epoch, as well as explaining where water exists today on Mars and how it affects the Martian atmosphere and surface. Mellon also critically considers the exciting but difficult possibility of extraction of water from the Martian poles and crust for human colonies of the future.

Moving outward from Mars, through the asteroid belt, we come to many diverse sites of water in the outer solar system. The exotic geology of the icy moons of the giant planets is the subject of Schenk's chapter (Chapter 5). The Galileo orbiter mission around Jupiter found very strong evidence for a possible liquid water ocean beneath the icy crust of the moon Europa, the smallest of Jupiter's four giant galilean satellites.

Out in the regions far from the Sun we also find ices of other molecules. Ammonia, methane, and carbon dioxide may have been incorporated in some moons, particularly those of Saturn, Uranus, and Neptune. Saturn's galilean-sized moon Titan appears to have acquired enough methane so that it now possesses a complex, non-biological organic chemical cycle between its thick atmosphere and haze-shrouded surface. This frigid moon, bigger than Mercury but likely composed at least half of water ice, may teach us something of the chemical steps leading to the formation of life. The US–European joint Cassini–Huygens mission began its trip to the saturnian system in October 1997, and will tell us more about this mysterious moon, the smaller icy moons, and the beautiful rings of Saturn, which at least in part are composed of water ice.

At the fringes of the solar system temperatures are so low that water ice is like rock, and ices of other molecules dominate. In Chapter 6, Stansberry describes the bizarre worlds Triton (a moon of Neptune) and Pluto. Both contain vast deposits of nitrogen, methane, and carbon monoxide ices, and both worlds have climates that are characterized by the seasonal cycling of these exotic ices into atmospheres and back on to their surfaces. Pluto and its moon Charon both contain water ice on their surfaces, and Triton presumably does too, though there the water may be hidden by a more prodigious inventory of other ices.

Finally, we come to the Kuiper Belt objects and comets, relics of the early stages of assembly of the planets from tiny grains. Discovering water ice on these small objects is surprisingly difficult, as Cruikshank describes in Chapter 7, but new technologies applied on progressively larger telescopes have brought success to the search. The identification of many other ices and organic molecules on comets, and now Kuiper Belt bodies, lends credence to the notion that Earth and its sister planets inherited these molecules (at least in part) from the deep outer solar system.

The search for the presence and nature of water ice throughout the solar system is an effort largely enabled by humankind's newfound talent for the exploration of the solar system. As with all such endeavors, planetary exploration has had its fits and starts, but the current generation of robotic explorers are remarkably diverse and capable. Their findings and promise of discoveries yet to come are suggestive of a future in which we will rely increasingly on direct sampling and analysis to understand the nature and origin of solar system ices.

From the point of view of a dispassionate extraterrestrial viewing our solar system, water ice is far more abundant and far more pervasive in its effects than liquid water. In fact, other than the water clouds of the giant planets and the putative ocean beneath Europa's surface, only Earth bears liquid water at present. And yet it is liquid water that provides the universal solvent without which life as we understand it could not exist. It is the delicate balance between atmospheric evolution and geological (and perhaps biological) processes that has maintained a terrestrial climate equable for liquid water over almost all of Earth's history. There is therefore much magic in the subtle difference between water as liquid and water as ice: without the former there might be no intelligence to contemplate the latter's importance in the solar system.

ROBERT BINDSCHADLER
NASA Goddard Space Flight Center

The history and significance of ice on Earth

We know more about the ice on Earth than on any other planet. Depending on where we live and the severity of our winters, many of us have familiarity with ice or its softer sibling, snow. Some of us play on it, while others curse the hardships it can bring. Caught up in these personal contacts, we rarely stop to appreciate its special properties, what its presence on this planet has meant for the habitability of Earth, and how it continues to affect the shape and size of the continents upon which we live.

Snow usually is delivered to us on the wings of chilling winter storms. Most rain begins as snow at great altitudes. Yet the majority of terrestrial ice occurs beyond our sight, as ice sheets and glaciers, confined to high latitudes of the planet and to its higher elevations. These remote, frozen reservoirs hold nearly 80% of the freshwater on Earth. While glaciers may seem plentiful, the vast Antarctic ice sheet contains nearly 90% of this ice, the Greenland ice sheet another 9%, leaving the smaller ice caps and glaciers throughout the world with only the remaining 1% (Fig. 1.1).

Other chapters of this book describe ices of exotic compositions that can be found elsewhere in the solar system. How boring then that on Earth we can only experience water ice! Not to be outdone by our interplanetary neighbors, our "water planet" contains water ice that manifests itself in a dazzling array of forms. This is most true at the small scale of single snowflakes. The hexagonal (or six-sided symmetry) of the ice crystals leads to a variety of snowflake forms (needles, plates, and the well-known dendrites) depending on the precise temperature conditions at the formation site. With near limitless possibilities of the conditions for formation, each snowflake ever created has been unique.

Nature's icy smorgasboard extends to larger scales, too. Extensive sheets of frozen ocean can be razor thin to many kilometers thick. The pattern of fractures and raftings of these sheets is constantly changing. Even the thicker, kilometers-thick ice sheets defeat any attempt at singular characterizations. Some move rapidly, others slowly; some receive a great deal of precipitation, while others are starved for snowy nourishment; some are windy, polar deserts where shifting snow replaces

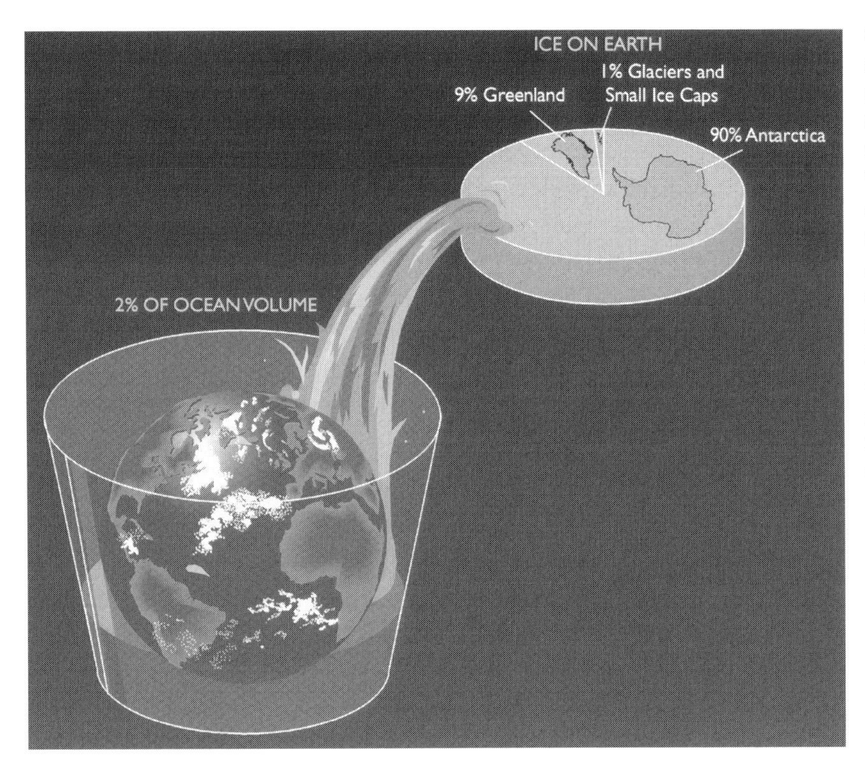

Figure 1.1 The present volume of ice on Earth accounts for 2% of water in any form. This fraction has grown to be more than 5% during the cold "glacial" periods. The ice is disproportionately shared between hemispheres with the Antarctic ice sheet accounting for 90% of the ice mass, the Greenland ice sheet another 9%, leaving the remaining 1% distributed elsewhere across the planet as smaller ice caps and mountain glaciers.
Credit: R. Bindschadler/NASA.

sand in the ever-changing barren landscapes, while others are deadly calm, permitting the formation of sparklingly large lacy single crystals. Most are bitterly cold, yet some glaciers extend into very warm climates. Indeed, ice can be found on all continents except Australia, dominating the polar regions and nearly touching the equator in Africa, South America, and even Indonesia.

Ice can be studied in many ways that lead us to learn more about the climate of our planet. An understanding of terrestrial ice provides a basis of understanding from which to embark on investigations of other planets in our solar system.

The water planet

Hydrogen and oxygen are abundant on the Earth (Toby Owen explains why in Chapter 3). The abundance of these elements, along with the temperature resulting from our distance from the Sun – conveniently warmed an extra amount by greenhouse gases in our atmosphere – allow water and ice to occur in ample supply, making Earth the "water planet." Earth's oceans provide a huge water source from which the ice sheets draw and return H_2O. The amount of gaseous water vapor in the atmosphere is extremely small by comparison. The total amount of water on Earth has remained relatively fixed.

Water defines the color of the planet, prescribes the shapes of the continents, and makes this world immanently habitable. The particular thermal conditions attributable to the presence of so much water have only recently been attained, but as life on this planet was establishing itself, so the ice on the planet was establishing a record of growth and decay.

Early glaciations

Geologists have found no evidence for the presence of ice during the first half of the Earth's 4.6 billion year history. This, despite the fact that the Sun's lower luminosity would have warmed the Earth 10%–20% less than today. The compensating influences were probably an atmosphere exceedingly rich in carbon dioxide and other greenhouse gases, capable of warming the air an additional 20 K, and younger, thinner, and warmer tectonic plates heated by the Earth's molten interior.

The first glaciation event occurred 2.3 billion years ago. It is suggested that a sudden increase in photosynthesis activity caused by biota bursting forth on the aquatic scene drained massive amounts of carbon dioxide from the atmosphere greatly reducing the greenhouse effect, cooling the planet. Subsequent to this climatic event, there is a puzzling period of 1.4 billion years (up to 900 million years ago) when no further glaciations took place. Some paleoclimatologists believe that erosion has simply eliminated the glaciation record over this interval. Others believe ice formation was successfully discouraged by the presence of thin, warm tectonic plates devoid of the higher elevations where ice preferentially initiates and suppresses the equator-to-pole transport of heat that effectively cools the polar regions.

At least six more distinct glaciations occurred during the next 800 million years. The first four were of continental scale between 900 million and 500 million years ago. The evidence is contained in the chemical composition of the rocks that were formed during this period. The last two glaciations, during the last 500 million years, were milder, taking place in the midst of generally warm climates. These two glaciations occurred at ancient high latitudes as the continental plates were locked in their Gondwanaland configuration. Even the Sahara basin took its turn in the polar regions and was covered by an extensive ice sheet. At the end of this geological period, the plates forming Gondwanaland rearranged themselves to form Pangea.

The last 100 million years have been marked by the slow dispersal of Pangea – a process continuing today as oceans widen, mountains rise, and plate boundaries grind and shudder. Reconstructions of paleoclimatic conditions improve dramatically over this last 100 million years of history because the records in ocean basins are intact, if growing, and less erosion of the terrestrial record has occurred. Oceanic circulation appears to have stabilized some tens of millions of years ago, leading to the formation of the Antarctic ice sheet 12–14 million years ago. Northern hemisphere glaciation was initiated several million years later. These ice sheets are still with us today; however they have fluctuated in size by significant amounts.

Ice sheets, sea level, and climate

Presently, only 2% of the Earth's water exists in frozen form. However, our present climate is as warm as any Earth has experienced for the past million years. Thus, this 2% figure is a minimum (at least for the past few million years). Climatologists call these warm periods "interglacials" because they occur between colder "glacials" when ice sheets and glaciers are more extensive. The last million years clearly show the oscillatory behavior from glacial to interglacial and back again. While ice sheets waxed and waned, sea level fell and rose in lockstep, first exposing and then submerging coastal areas of all the continents. Over the past million years, sea level has risen and fallen over 125 meters (Fig. 1.2).

The clearest view of a glacial period that paleoclimatologists have pieced together is of the most recent glacial period that ended just 20,000 years ago. It also appears to have been one of the more extreme glacial periods. Large ice sheets, up to 5 kilometers thick draped themselves over the northern halves of Europe and North America (Fig. 1.3). In mountainous areas, these frozen mantles extended icy white fingers

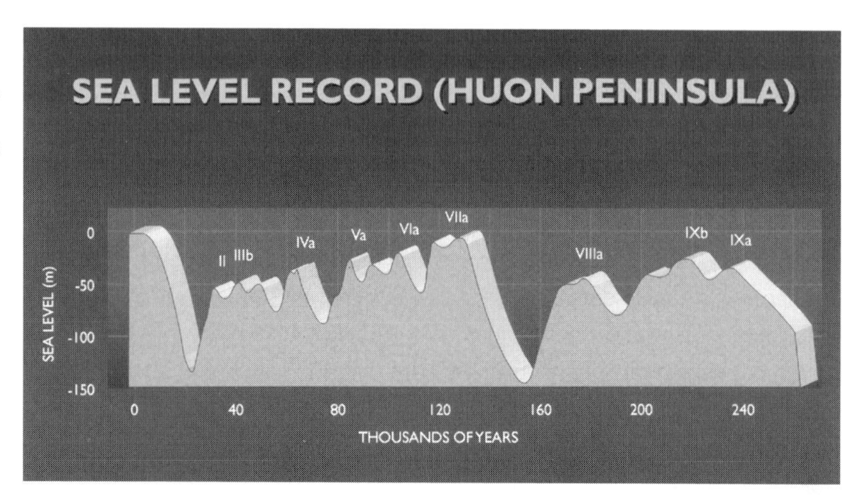

Figure 1.2 In the last quarter million years sea level has oscillated more than 150 meters. Our current warm interglacial period is extreme relative to this recent history and coincides with a high stand of the oceans. *Credit:* R. Bindschadler/NASA.

Figure 1.3 Northern Hemisphere ice sheets were far more extensive during cold glacial periods. This illustration depicts the pervasiveness of the ice sheets during the last glacial maximum 20,000 years ago. *Credit:* R. Bindschadler/NASA.

Figure 1.4 Yosemite Valley is an excellent example of U-shaped valleys that are formed by the erosion of glaciers against the land surface. This classical shape of glaciated terrain results from the internal motion of the glacier with fastest motion, thus greater erosive force at the center and relatively less motion at the glacier margins.
Credit: Public Domain illustration.

into warmer regions carving out classic U-shaped valleys by the slow, erosive motion of the heavy ice against the rock (Fig. 1.4).

During even the most extreme glacial periods, the Antarctic ice sheet was unable to expand significantly because its present perimeter lies close to the edge of the continental shelf, beyond which the ice sheet looses its grip on the underlying bed. Stormy seas and violent winds fragment any ice venturing beyond this continental limit into icebergs.

In total, the ice sheets of the last glacial period drew enough water out of the oceans to convert the coastlines of the world into an unrecognizable configuration (Fig. 1.5). During this maximum, 5% of the world's water was contained in ice sheets, two and a half times more than we see today, and as large as it has been in the last million years. With less water, more exposed land and winds sweeping down from the vast ice sheets, the weather of glacial periods was not only colder, but windier, dustier, and drier as well.

Our current interglacial period has been interrupted at least twice. Once by a brief return to full glacial conditions (cold and dry) 11,700 years ago, that may have only affected the Northern Hemisphere, not lasting long enough to chill the Southern Hemisphere into a companion cold spell. Analysis of recent ice cores from Greenland show that the switch into this glacial interlude happened in fewer than three

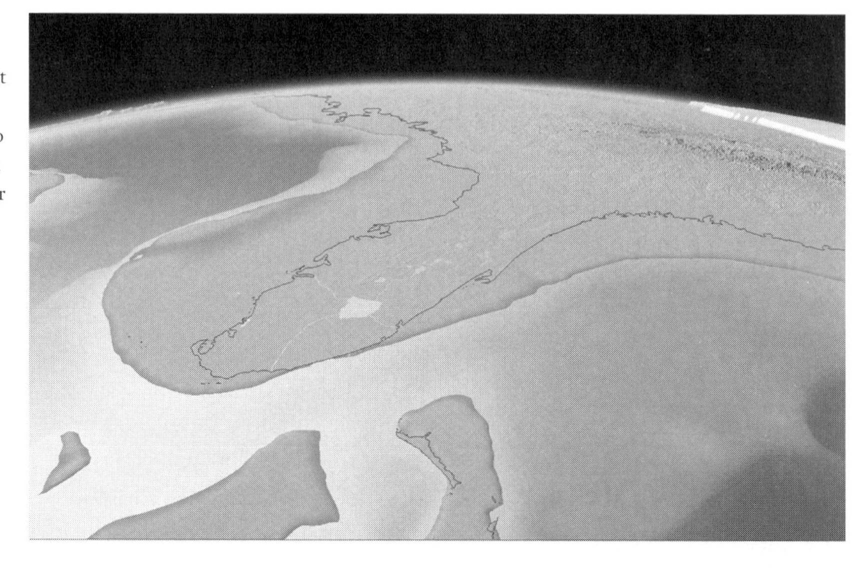

Figure 1.5 The position of the coastline in the vicinity of the Florida peninsula during the last glacial maximum (upper figure) (see the ice sheet impinging into the view on the right) and if sea level were to rise 5 meters (lower figure). The flat coastal plains comprising the continental shelves are susceptible to large shifts in coastline position caused by relatively minor increases or decreases in sea level.
Credit: W. Haxby/LDGO.

years – a climatic "blink of an eye." The less extreme cold period is the more familiar Little Ice Age, which peaked in the mid 1730s and caused extensive abandonment of farms in northwestern Europe as well as severe human problems elsewhere around the world.

Surprisingly, the average temperature difference between glacial and interglacial periods is only 3 K. Most climatologists believe that regular variations in the Earth's orbit act as the pacemakers driving the recent glacial–interglacial cycling. The celestial direction of the Earth's pole wobbles at a period of 19,000 and 26,000 years. Also, the non-circularity of Earth's orbit around the Sun varies at a 41,000 year cycle. These effects combine to match closely the recorded variations in the climate. A paradigm that remains is that these variations in the amount and seasonal timing of the Sun's warming rays cannot account for even the few degrees difference between climatic extremes. The source or sources of the thermal amplification within the climate system remain mysterious, but are actively pursued by today's climatologists.

One candidate process that leads to such a climate amplification feedback involves the interaction of ice sheets and sea level directly. Warmer temperatures lead to a gradual reduction in ice volume through increasing melting, causing sea level to rise. As sea level rises, the large grounded ice sheets that terminate in water are forced to retreat. When this happens, the thinning effect propagates upstream, releasing ice into the ocean from areas that never melt. This discharge further raises sea level, causing more retreat and more thinning. It is believed by some that this effect is responsible for the phased demise, beginning 20,000 years ago, of the Northern Hemisphere ice sheets (which raised sea level initially) followed by the reduction of the Antarctic ice sheet at the termination of the last glacial cycle and the beginning of our current interglacial interlude.

Expectations of a reversal to colder conditions as dictated by the future geometry of our celestial orbit fly in the face of predictions endorsed by most climatologists that anthropogenic activity will continue to warm the planet. As scientists debate the issues, human, natural, and interplanetary forces are engaged in a struggle of genuinely global proportions, driving us into uncharted climatic territory for which there is no recent, well-documented historical analog. It may prove unfortunate that shorelines have been relatively stable for the last few thousand years, lulling us into a false sense of security that sea level will not change rapidly. Nearly half of the world's population now lives within a few miles of a coastline. The historical record strongly suggests that this stability will not last indefinitely, possibly leading to an eventual migration of billions of people.

Ice sheet response

Ice sheets are cold, but they are not static. Their size and shape adjust to climate in such a way that ice flow will transport snow accumulating

Note 1.1 Flow inside a glacier (or ice sheet)

In general, more snow accumulates at higher elevations. Thus, the winter snowpack is thicker in the upper areas of a glacier. In contrast, melting, driven primarily by summer temperatures, is larger at lower elevations. This leads to a net excess of mass added each year at higher elevations and a net mass deficit at lower elevations. If ice did not flow, the net gains and losses would accrue, steepening the profile of the glacier (Fig. 1.6).

The flow of the glacier is determined through the glacier's shape (and temperature). There is an interdependence of geometry and flow rate so that adjustments to one result in adjustments to the other. Under constant climate conditions, adjustments continue until they each support a configuration that exactly compensates for the spatial pattern of mass gains and losses across the glacier. Thus, in the accumulation area, where mass gain is realized annually, the ice velocity will be downward relative to the surface by an amount that is proportional to the net mass gain. Correspondingly, in the ablation area, where ice is lost annually, ice will arrive from upstream by glacier flow in the amount that matches the ice lost to melting and other meteorological processes. The flow direction in the ablation area will be upward relative to the surface. The overall result will be no change to the surface profile.

It can be seen from this simple picture that ice which accumulates in the highest regions of the glacier will flow slowest, traverse deepest, and emerge farther downstream than ice that accumulates at a lower elevation.

If climate remains stationary for long enough, the glacier will be able to achieve this equilibrium configuration. The response times of glaciers and ice sheets depends on their size and flow rates. For small, fast glaciers, the response time can be as short as a decade. For the largest ice sheets, response times are much longer – tens to hundreds of thousands of years.

in the higher, colder regions of an area to replace ice lost by melting at warmer, lower elevations. As atmospheric temperatures change, the changes affect ice sheets both by altering snow accumulation rates and by changing the temperature of the ice. Changes in accumulation and melting affect ice sheet volume immediately. It has been confirmed that recent atmospheric warming in the Antarctic region has been accompanied by increases in snow accumulation. This relationship holds because warmer air is able to hold and transport more moisture to the ice, where it falls as snow.

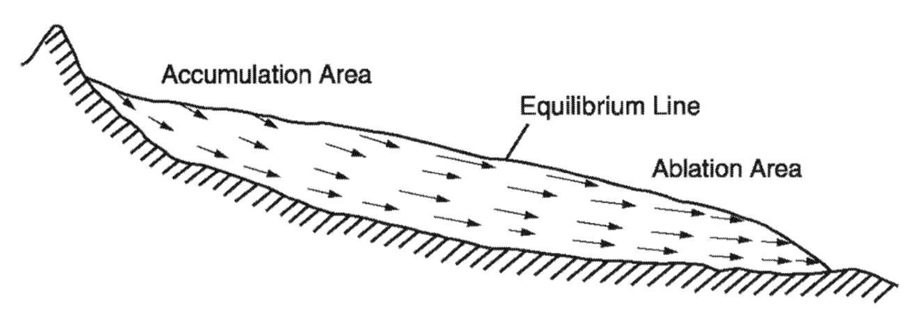

Figure 1.6 Motion of the glacier redistributes mass to compensate for net accumulation at the higher elevations and net melting at lower elevations. Motion within the glacier is largest nearer the surface and less near the head and terminus. *Credit:* adapted from W. S. B. Paterson.

Internal changes also take place within the ice sheet as atmospheric temperatures change, but these changes take tens of thousands of years to manifest themselves. Warmer ice deforms more readily than colder ice. Thus, as atmospheric warmth diffuses down into an ice sheet, the ice sheet can flow more quickly and needs less size to accomplish its task of shedding excess mass in its upper reaches and replacing mass downstream. As a result, the ice sheet becomes thinner. Because the conduction of temperature occurs slowly and internally, this response can be asynchronous with the climate trend at any particular moment, yet the larger, longer duration climate events will be expressed by seemingly synchronous changes in ice volume. Records of climate history and sea-level history confirm this.

Snow and sea ice

As cooler temperatures lead to an increase in ice formation, the high reflectivity, called albedo, of the ice surface returns a greater proportion of the incident sunlight back into space, further cooling the planet. This is another amplification effect contained in our climate and is referred to as the "ice-albedo" feedback. Ice and snow dominate this feedback mechanism because there are few natural surfaces as reflective as ice and snow. Fresh snow is the brightest of all surfaces at visible wavelengths.

The Earth's axis is tilted with respect to the ecliptic. As a result, temperatures undergo a strong seasonality. Winters are cold enough over much of the planet to create and retain an extensive winter snow cover. This bright snow cover causes a further reduction in the amount of solar radiation absorbed by the surface amplifying the seasonal cooling of wintertime portions of the planet and creating an additional

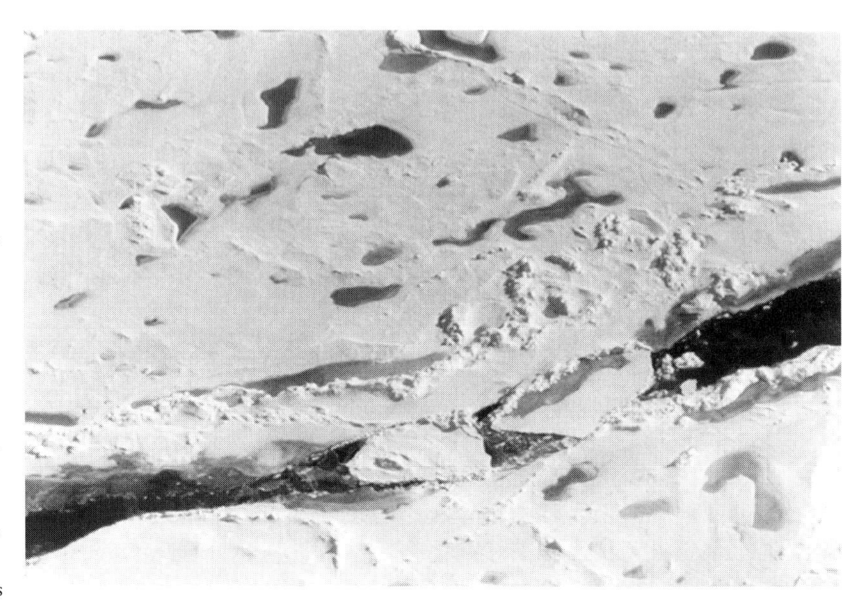

Figure 1.7 This aerial photograph of Arctic sea ice shows many of the different structures that are formed by the constant movement of the ice and the response to solar radiation. The narrow, darkest region is an open lead where ice floes have separated, exposing the ocean. In the gray areas within the open lead thin ice has begun to form where the surface waters are exposed to the cold air. The brighter ice consists of a number of ice floes. As these floes move, they crash into each other piling up colliding ice into ridges that can be seen as areas of jumbled ice blocks. The darker rounded areas are melt-ponds, formed by melted surface snow contained within surrounding ice.
Credit: photo by C. Parkinson/NASA.

positive feedback that makes our climate more sensitive to changes in temperature.

This amplified seasonal cooling is not confined to the land. It occurs within the high latitude oceans as well. During the continuous darkness of the polar winter, ocean water becomes so cold that it, too, freezes creating sea ice. Sea ice, and the snow that may fall on it, further extend the area over which a highly reflective surface greets the returning Sun and delays the summer warming.

Ocean currents and surface winds drag across the rough surface of the sea ice and move it. Because the density of ice is nine tenths that of water, most of the floating ice exists below the waterline and that roughness of the underside of the pack is nine times more extreme than the surface roughness. Nevertheless, the surface winds are much faster than the ocean currents. On balance the wind and the currents each contribute roughly equal amounts to the motion of the ice pack. Often when it moves, it fractures. The differential motion of adjacent pieces, or floes, can lead to either a piling up, forming ridges (Fig. 1.7), or the opening of gaps, called leads, in the ice cover. If cold conditions persist, these leads rapidly freeze over, increasing the total amount of sea ice.

Formation of sea ice provides a number of influences on the ocean from which it came. Sea ice is not very thick by ice sheet standards – 3 meters or less, but this is thick enough to cut off most of the heat exchange between the relatively warm ocean and relatively cold polar atmosphere. Impediments to the ocean-to-atmosphere heat exchange amplify temperature differences in the atmosphere between the polar regions and the tropics.

When sea ice forms, salt is not incorporated into the frozen matrix but remains in the water. This increases the water's salinity in the layer just beneath the forming sea ice. Saltier water is denser. This denser water sinks, helping to stir the upper layers of the ocean during winter's freezing season. In summer, when the fresher ice melts, the upper layer of the ocean receives a surface layer of fresh water, stabilizing the warm column.

As the seasons shift, the snow and sea ice covers shift from Northern Hemisphere to Southern Hemisphere and back again. Although the amount of sea ice in each hemisphere varies by a factor of 10 or more between summer and winter, the global total varies less than 50%. With considerably more land mass in the Northern Hemisphere than in the Southern Hemisphere, the snow cover effect is more asymmetric – the global snow cover blanket being much larger from December to February than from June to August.

Figure 1.8 Polar views of the Northern and Southern Hemispheres with elevations coded in white through deep gray tones. Ice sheets are among the highest features of the planet. This bipolar view shows the elevation of the Earth's surface and the bathymetry of the oceans. Darkest hues are below sea level. Higher elevations are shown in progressively lighter tones. Only the Himalayan range competes with the high-standing ice sheets in Greenland and Antarctica. *Credit:* NOAA.

Ice sheets and weather

Quite distinct from the strongly seasonal components of the cryosphere, the ice sheets are perennial, supplying the large heat sink that – along with the heat received preferentially in the tropics – drive the longer-term global-scale equator-to-pole atmospheric circulation. This temperature contrast is similar in each hemisphere, yet there are important differences in the arrangements of the ice sheets in each hemisphere that produce significantly different circulations within the ocean and atmosphere (Fig. 1.8).

The Antarctic ice sheet is roughly centered on the South Pole and extends more than 2,000 kilometers in all directions. Around it, the southern portions of all the oceans are connected, allowing a persistent circumpolar current to be established and maintained, forcing the atmospheric circulation carrying individual storm centers to swirl around the perimeter of the continent in a west-to-east, or zonal pattern similar to that of the ocean. This combination of natural forces results in the "roaring forties" and "raging fifties" feared by mariners who traversed this region and is responsible for extended periods of seasickness for even the saltiest sailors.

This massive ice sheet also ascends to a greater average height than any other continent. The South Pole's elevation is more than 2,800 meters above sea level. Much of the remainder of the ice sheet is even higher. The extreme cold and considerable height of the ice sheet combine to limit the amount of moisture storms can deliver as well as limiting the ability of the storms to penetrate into the elevated interior regions of the ice sheet. The result is a high-elevation polar desert where less than 2.5–5 centimeters of snow accumulate all year. Along the continental perimeter, however, the accumulation rates are much larger. This topographic barrier, a result of the parabolic shape of the ice sheet, not only forces the storms to release their snow along the perimeter of the ice sheet, but generates a persistent cloud cover for which the coast of Antarctica is famous.

In significant contrast to Antarctica, the Greenland ice sheet lacks a pole-centered symmetry, but, instead, stretches for 2,000 kilometers south to 60 degrees north latitude. By rising up to elevations similar to those in Antarctica, of more than 3,200 meters, this ice sheet blocks the establishment of a strong zonal circulation in the Northern Hemisphere. Unable to efficiently cross this barrier, air masses deflect around it. Disruption of this zonal flow causes the southward trajectory of weather patterns from Canada to the United States and contributes to a more variable pattern of weather in the Northern Hemisphere.

The accumulation pattern in Greenland is strongly influenced by this blocking characteristic. Most of the accumulation is delivered to the western side of the ice sheet because the mean zonal flow at these latitudes is from west to east. The eastern side of the ice sheet is precipitation deprived. As a result, the ice sheet's western side is more extensive and there is a noticeable eastward shift in the position of the ice divide.

The ice sheets are so large they contribute to their own weather pattern. As snow accumulates on their surface they rise to higher and, therefore, cooler temperatures, helping to preserve their mass.

Note 1.2 Shape of an ice sheet

Gravity makes ice flow. Like many materials, ice deforms when forces are applied to it. The larger the force, the greater the rate of deformation. At the molecular scale, the deformation process depends on the motion of imperfections within the crystal structure. These imperfections move more rapidly at warmer temperatures making warm ice "softer" than cold ice. Water ice on Earth has a crystalline structure that contains internal planes along which deformation occurs preferentially. The reactive deformation is non-linear, so much so that a worthwhile approximation is that ice behaves as a plastic material. "Plastic," to a materials scientist, means that there is a threshold stress, the yield stress, below which no deformation occurs and above which deformation is infinite. With this approximation, the forces everywhere within the ice must be at, or below, the yield stress.

Forces reach their maximum at the base of an ice sheet. By evaluating the forces on a column of ice, the along-flow profile of an ice sheet can be determined. The force, F, acting on each vertical face of the column is the average weight of the ice times the face area

$$F = \frac{1}{2}\rho g H^2 W$$

where ρ is the density of ice (917 kg/m^3), g is gravitational acceleration (980 m/s^2), H is the ice thickness, and W is the face width (Fig. 1.9). The sum of these forces is balanced by a shear force at the base

$$F_b = \tau \Delta x W$$

where F_b is the force, τ is the yield stress and Δx is the horizontal length of the column. The uphill and downhill forces on the vertical faces oppose each other and their difference is equal to shear force along the base

$$\tau \Delta x W = \frac{1}{2}\rho g W\left(H_u^2 - H_d^2\right)$$

which is approximated by

$$\tau \Delta x = \rho g H \,\Delta H$$

Integration of this relationship determines the classic parabolic profile of an ice sheet

$$x = \frac{\rho g}{2\tau} H^2$$

For comparison, the shape of a cross-section of the Antarctic ice sheet is shown (Fig. 1.9).

Ice on a different planet would have a different aspect ratio between extent (x) and depth (H) owing to a different value of g. In addition, this equation can predict the aspect for an ice sheet of a different material by substituting the appropriate values of the yield stress and density.

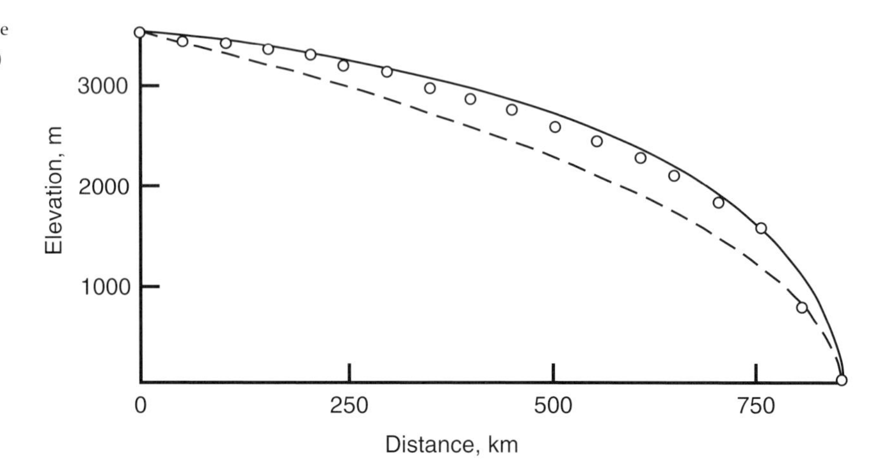

Figure 1.9 A cross-section of the East Antarctic ice sheet (circles) can be approximated by assuming ice acts as a plastic material that produces a parabolic profile (dashed line). However, the fit is improved if the theory treats ice as a non-linear viscoelastic solid (solid line).
Credit: adapted from W. S. B. Paterson.

Increasing their elevation also contributes to the extraction of additional precipitation. These positive feedbacks only operate to a point, however. Eventually, the ability to deliver moisture decreases with temperature, so colder temperatures begin to become a losing proposition and fail to result in additional precipitation. Over thousands of years, the ice sheets grow to a point where these effects attain a long-term balance.

Metamorphism of snow into ice

Although the ice sheets are fed primarily by snow, they are comprised primarily of ice. This transition, called metamorphism, occurs over time beneath the surface. Molecule by molecule, the fine, sharp-edged snow grains, which began as single snowflakes, change to more rounded forms. Molecules can move more rapidly at higher temperatures, so the rate of metamorphism is temperature dependent. Large snow grains grow at the expense of smaller grains. Liquid water greatly accelerates this process by providing a fluid vehicle to rearrange H_2O molecules.

The weight of additional overlying snow compacts the snow matrix adding to the densification process. The evolution toward more rounded snow grains accentuates this process. Eventually, air passages within the snow pack pinch shut, trapping bubbles of air. Further compression increases the pressure of the air bubbles. At a density of 917 kg/m^3 (water has a density of 1,000 kg/m^3), pure ice is formed, the

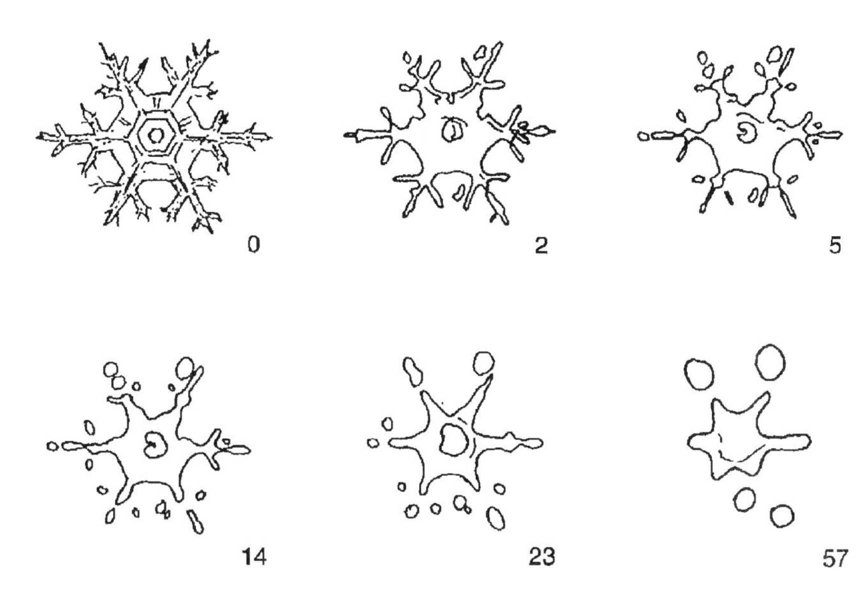

Figure 1.10 Over time individual snowflakes lose their lacy appearance and eventually form ice. Most of the transformation occurs over the first year or two. The most delicate features are lost within a few days as individual molecules move to less-exposed locations. As this process continues, the arms of the snowflake are lost as the central nucleus grows. Numbers indicate the age of the snow grains in hours.
Credit: adapted from E. LaChapelle.

ice crystal lattice is complete, and further compression ceases. Many exotic high-pressure forms of ice have been generated in the laboratory, but these do not occur naturally on Earth.

The depth of this metamorphosing layer of snow, called firn after it is one-year old, varies with the intensity of the processes that drive the densification. In warm, wet regions, such as the Alps, ice can be found ten meters below the surface of a glacier in its accumulation area. In the Antarctic polar desert, the transition to ice lies closer to 100 meters below the surface.

Climate tape recorder

One of the most exciting uses of ice has more to do with the air bubbles trapped within the ice than with the ice itself. The gradual metamorphism of snow into ice amounts to a tireless, non-human laboratory assistant patiently taking air samples and sealing them away in nature's freezer. All that scientists must do is exhume these pristine samples of ancient atmosphere to conduct precise chemical and microparticulate analyses on them (Fig. 1.11). On Earth, this research has led to startling results on the rapidity of climate change – temperatures changed in Greenland by 10 K in just a few years. Scientists are conducting similar ice-corings in Antarctica and finding equally rapid changes, confirming

Figure 1.11 Ice is extracted from the ice sheet by drilling a core in sections and lifting each section to the surface. The clarity of the ice depends on the number and size of air bubbles. Banding within the ice formed by seasonal weather differences and by volcanic events (that leave dust bands) help scientists date the age of the ice. Near the bottom of the ice sheet, the core can contain larger amounts of dirt that are incorporated into the ice by motion of the basal ice.
Credit: photo by K. Taylor.

that drastic climate changes can occur globally. In the deepest ice core yet drilled, in the interior of Antarctica at Russia's Vostok station, atmospheric samples for more than 400,000 years have been recovered.

This unique paleoclimatic approach has its limits. Of course, no atmospheric samples will be taken until the ice sheet begins to form, but, even then, the removal of ice from the bottom will allow the precious air samples to escape and mix. In an ironic reversal of fortunes, a thick ice sheet has the unwelcome property of insulating deeply buried ice from the cold atmosphere so efficiently that trapped geothermal heat can melt the lowest layers of the ice sheet erasing the precious paleoclimate record before it can be read.

However, this dark cloud has a silver lining. Underneath Vostok station, a Russian outpost situated on the surface of some of the thickest ice in Antarctica, there lies a lake the size of Lake Ontario. The water filling this lake is hundreds of thousands of millions of years old and has been isolated from any direct contact with the atmosphere for more than 10 million years. An exciting exploration event awaits scientists who are preparing ways of tapping into this ancient lake. It is a project complete with the unique engineering challenges and stimulating anticipation usually reserved for extraterrestrial research.

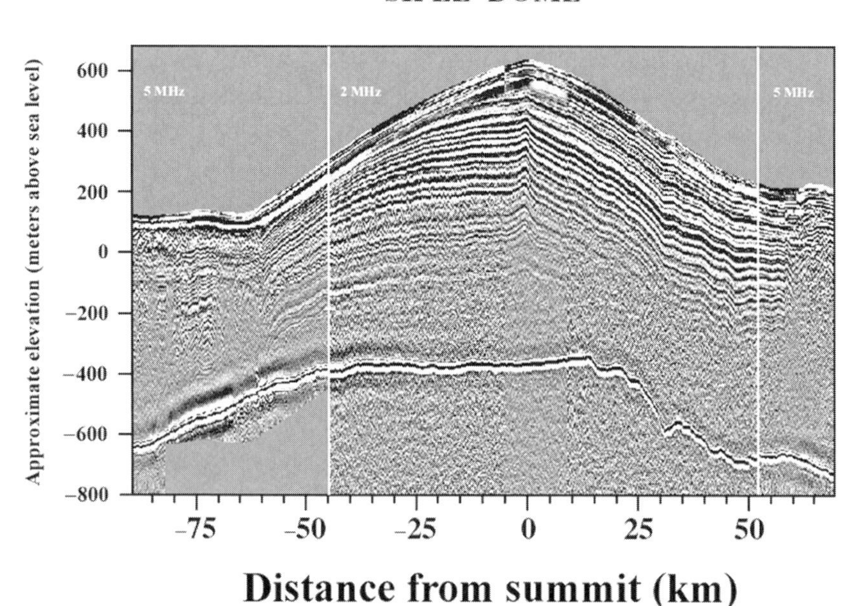

Figure 1.12 Radio waves transmitted into the ice produce echoes that are processed into detailed views of the internal structure of the ice sheet. In this example a local ice dome in West Antarctica is traversed with just such a radio echo sounder. The measurements show the thickness of ice reaches a maximum of 1,000 meters and rests on a bed (the strong line near the bottom of the figure) that is well below sea level. Internal layers correspond to isochrones (contours of equal age) and represent the current position of former surfaces of the ice sheet.
Credit: image by R. Jacobel.

Even without drilling through the ice, there are useful properties of the ice that can be sensed by penetrating radars. For glaciologists, keen on measuring the depth of the ice, the identification of additional layers within the ice has furthered their attempts to understand the ice sheet flow history (Fig. 1.12). These layers have been confirmed to correspond to slight contrasts in electrical conductivity that, fortuitously, also correspond to isochrones, or surfaces of equal age.

Ice sheet facies

The interaction of the seasonal accumulation and melting on ice sheets and glaciers leads to the development of a series of distinct diagenetic zones, or facies, that can help define the balance between these processes at any site. In the interior of the ice sheet, if conditions are cold enough that no melting occurs, even during mid summer, the metamorphism of snow into ice proceeds as described above. This is called the dry snow facies. At lower elevations, where melting takes place, in addition to accelerating metamorphism, this melt-water percolates downward into the cold snow pack where it refreezes into ice as a network of vertical pipes and horizontal lenses defining the region called the percolation facies.

A thermal consequence of this refreezing is the release of latent heat at the refreezing site in an amount equal to the heat absorbed at the surface to originally melt the snow. This process effectively transfers heat from the surface within the snow pack. If enough heat is released within the snow pack to raise its internal temperature to the melting point, no further refreezing can take place. Additional melt-water will be held within the snow pack until it reaches saturation (about 4% water) creating the slush facies, at which time further melt-water will percolate deeper or drain away horizontally. Eventually, so much melting takes place, that all the previous year's snow is removed. What is left is an ice surface, the bare ice facies. In some of the coldest regions, the edge of the ice sheet is reached before the warmer facies can develop. In these situations, mass is lost as calved icebergs from the ice sheet's edge.

These facies differ mostly by the nature of their subsurface characteristics. Optical or thermal imagery can discriminate some of these facies, but radar is more suited to this task. Radar imagery transmits a radar pulse at a frequency that penetrates snow, causing the character of the returned radar energy to depend in large measure on the subsurface structure where the distinctions between facies are most pronounced (Fig. 1.13).

Ice sheet motion

Ice seems solid and immutable; however it does respond – albeit slowly – to mechanical forces imposed on it. In the natural setting, gravity is the force responsible for moving ice sheets. Precisely, ice is a non-Newtonian visco-elastic solid, however the essential characteristic of ice sheet flow can be illustrated by approximating ice to be a plastic fluid. With this approximation, it can be shown that ice sheets will have parabolic cross-sections of ice thickness. More complex, but increasingly realistic approximations demonstrate that the motion is fastest at the surface, decreases with depth, and on an ice sheet will increase from the interior to the perimeter.

As discussed earlier, much of the significance of terrestrial ice lies in the fact that ice moves. Many of the surface features that are seen

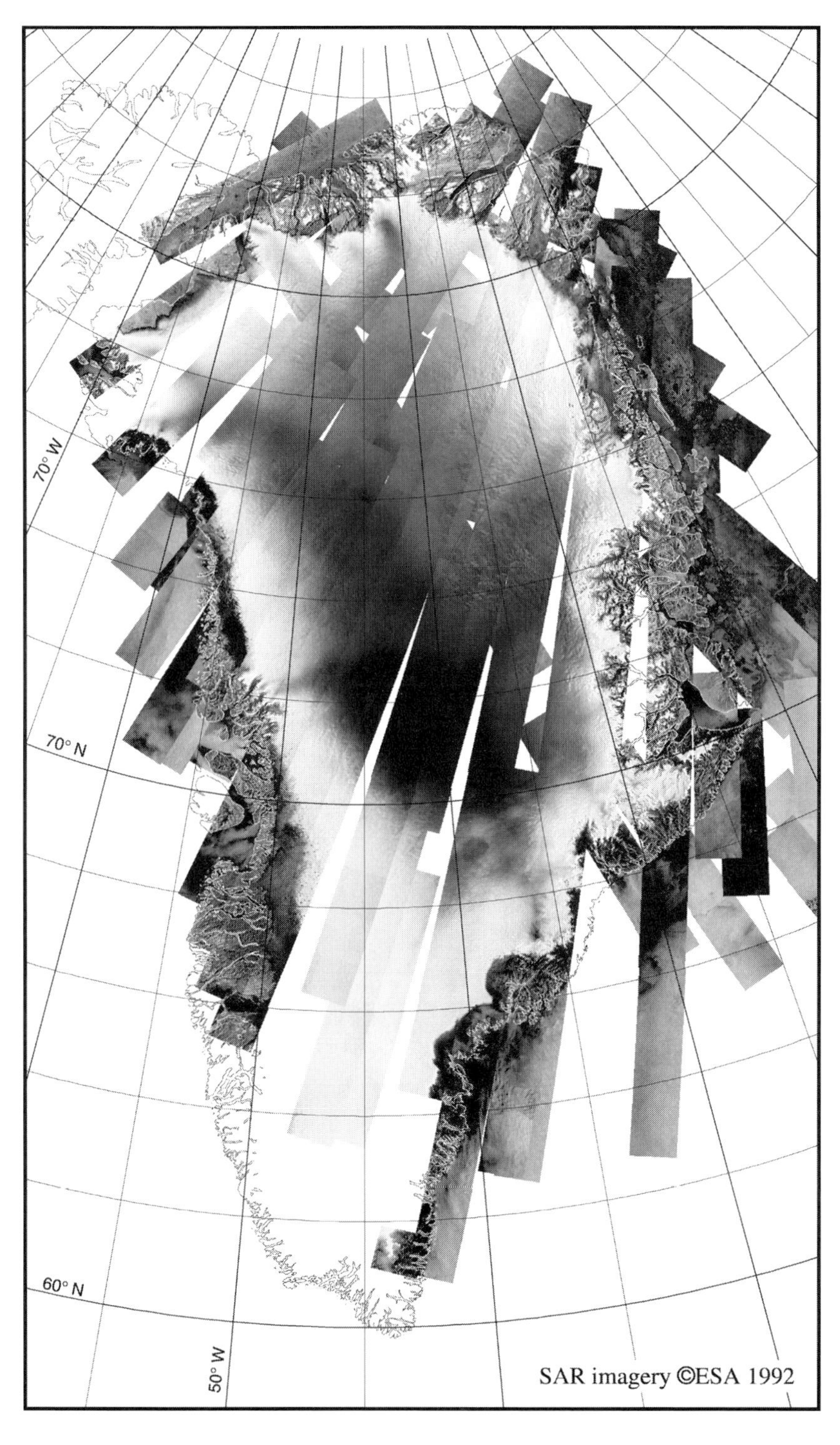

Figure 1.13 This mosaic of Greenland is composed of backscatter images collected by a spaceborne synthetic aperture radar. The central region of the ice sheet is dark because the radar's energy is absorbed by the dry snow at the highest elevations of the ice sheet. Nearer the coast, the backscatter increases due to internal scattering by refrozen melt-water layers within the percolation facies of the ice sheet. Along the margin of the ice sheet, the smooth ice surface (bare ice facies) scatters a smaller proportion of the transmitted radar energy.
Credit: data copyrighted by European Space Agency, image prepared by M. Fahnestock.

owe their form to the mechanical properties of ice. Because ice is not a good conductor of heat and snow is even poorer (this is one reason igloos work so well), ice sheets act as thick insulating blankets. The cold polar atmosphere and warm Earth interior combine to form a strong thermal contrast between the cold, stiff upper layers of ice and the warm, soft lower layers. The softness of the deeper ice experiences the greater weight of the ice above and deforms much more than the overlying ice. Motion at the surface is the result of the stiffer ice being carried along by the deformation at depth within the ice sheet.

A second process that sometimes occurs in ice motion is basal sliding. If the temperature of the basal ice is at the melting point, the ice in contact with the bed can move, lubricated by water at the icebed interface. In this case, the entire ice column moves as a unit and, again, the process is expressed by motion at the surface.

Were ice not to move at all, the surface of ice sheets and glaciers would be extremely smooth, much like a tranquil lake. However, the faster ice moves, the more the surface of the ice reflects the undulating character of the subglacial topography. This fact is a useful tool in the initial assessment of the dynamic character of an ice sheet.

When the ice motion is fast, compared with the rate at which surface bumps relax to a smoother shape, flowstripes develop (Fig. 1.14). The longevity of such features is a consequence of their minor influence on the stress field and the stiffness of the cold upper layers of the ice sheet (or glacier). Flowstripes show scientists which direction the ice is flowing. Misaligned with flow, they indicate that there has been a change in flow pattern.

Spatial variation in flow rates can lead to fractures of the ice, or crevasses. These highlight fast-flowing ice streams, rapidly moving rivers of ice contained within an otherwise sluggish ice sheet (Fig. 1.15). Crevasses have been used as convenient markers to track the speed and direction of these galloping glaciers (Fig. 1.16). Many ice streams move 1–2 meters per day. The fastest moving ice in Antarctica is the Pine Island Glacier that races along at 8 meters per day. Greenland holds the record, however, where the Jokobshavn Isbrae charges into the sea at over 6 kilometers each year.

Ice streams return the discussion to the issue of sea level rise. It is through the discharge of fast ice streams that the ice sheet can drain most rapidly, adding water to ocean basins and raising sea level. Recent

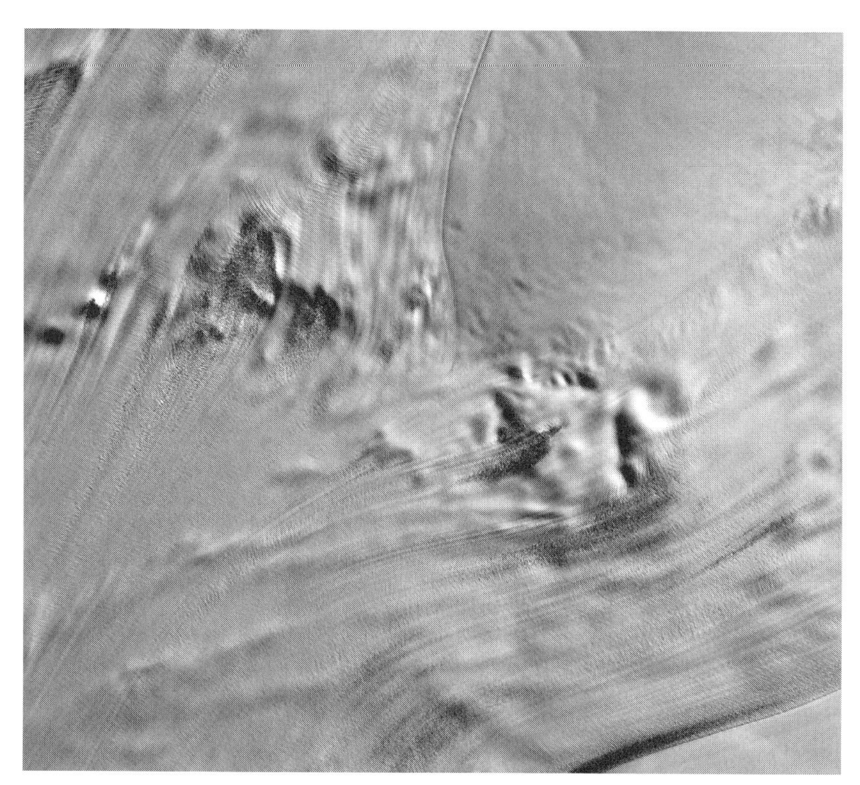

Figure 1.14 Flow at this confluence of two major West Antarctic ice streams is evidenced by the trajectories of flowstripes, formed by the moving ice. Fields of crevasses also tell of regions resisting flow where stresses on the ice increase, forcing the ice to divert and fracture. This Landsat image covers an area 85 km × 95 km. *Credit:* R. Bindschadler/NASA.

Figure 1.15 Aerial photograph of crevasses at the margin of an ice stream. Crevasses are cracks in the ice formed by the intense stress that exceeds the material's ability to stretch without fracturing. In this case the stress is created by the large difference in speed between the fast-moving ice stream and the slower ice adjacent to the ice stream. *Credit:* photo by N. Nereson.

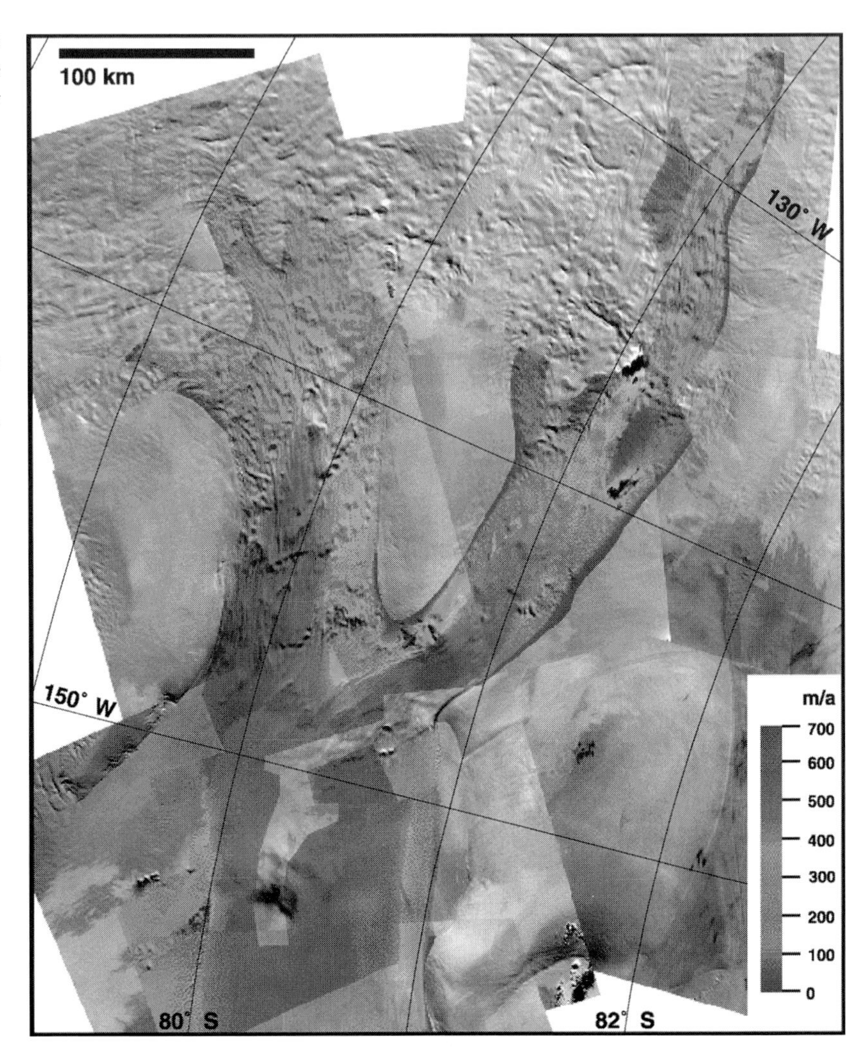

Figure 1.16 Colors represent the ice speed on these two major ice streams in West Antarctica. Flow is from top to bottom in the figure. Colors are superimposed on a mosaic of ten Landsat Thematic Mapper images that show the surface topography variations. The data show a general increase in speed until the ice streams join. At this point they enter the floating ice shelf where the lateral expansion of the ice results in a gradual deceleration.
Credit: R. Bindschadler/NASA.

research has shown that the speed of these ice streams is resisted by a combination of friction with colder adjacent ice and friction at their bases. By moving faster, they can generate more friction that could allow them to either widen or lubricate their beds more – both processes possibly leading to an increase in discharge.

Ice shelves

The enormous weight of ice sheets loads the crust enough to depress the crust many hundreds of meters. This is enough to place the bed of the ice sheet below sea level. Many of the ice streams and outlet glaciers exit the ice sheet across submarine beds to feed large floating ice shelves. In some areas, such as West Antarctica, the bed is submarine, even without ice loading, so this extra weight only makes the bed deeper allowing thicker ice streams to exit the grounded margin. Distinctly different from sea ice, floating shelves are hundreds of meters thick and flow by virtue of their considerable weight. Most exist in confining bays that provide protection from tidal and other oceanic forces that work to erode the exposed edges of the shelves through melting and ablation, fragmenting the shelves into icebergs. Icebergs are created in a wide variety of sizes. The largest are equal in area to some of the smaller US states (Fig. 1.17).

Friction along the interior perimeter of the shelves helps to support the shelves. Additional support for the shelf comes from areas within the shelf where the shelf runs aground. If the ice continues to move over the submarine bed, the area of disturbed ice is called ice rumples, whereas, if the grounding is more complete and the ice stagnates locally, the ice shelf must flow around the feature called an ice rise.

Ice landscaping

Ice motion is responsible for much of the landscaping of the Earth's surface. During the last ice age, glaciers and ice caps expanded to cover six times more area than they do today, leaving their mark on much of the land people now inhabit. This landscaping is a combination of depositional features, such as moraines, drumlins, and eskers left upon the surface as the ice melted; and erosive features, such as the glacial valleys themselves and striated bedrock, formed by mechanical force of rocks held by the ice and water flowing underneath the ice against the subglacial bed.

Much of the erosion is accomplished by rocks held within the basal ice being ground against the bedrock. Striations are formed in this manner. A critical fact that accentuates a glacier's erosive capability is that as water freezes, it gets less dense and expands. This process can rapidly make large cracks from small ones and break apart rocks subglacially. These unique forms of landscaping are often more easily recognized from space (Fig. 1.18).

Figure 1.17 Icebergs form when ice calves off the margin of the ice sheet. Massive floating ice shelves give birth to large tabular bergs that can be many miles on a side. Eventually wave action and ocean heat weaken the largest bergs forcing them to break into smaller pieces. Ice is nearly as dense as water leaving nearly 90% of the iceberg's volume submerged.
Credit: Public Domain illustration.

Meteorite catchers

While the ice sheets have been patiently sampling and preserving ancient atmospheres, they also have been catching space debris. An ice sheet acts as a snowy pillow protecting meteorites from disintegrating upon impact with Earth. Where the normal seaward flow of the ice

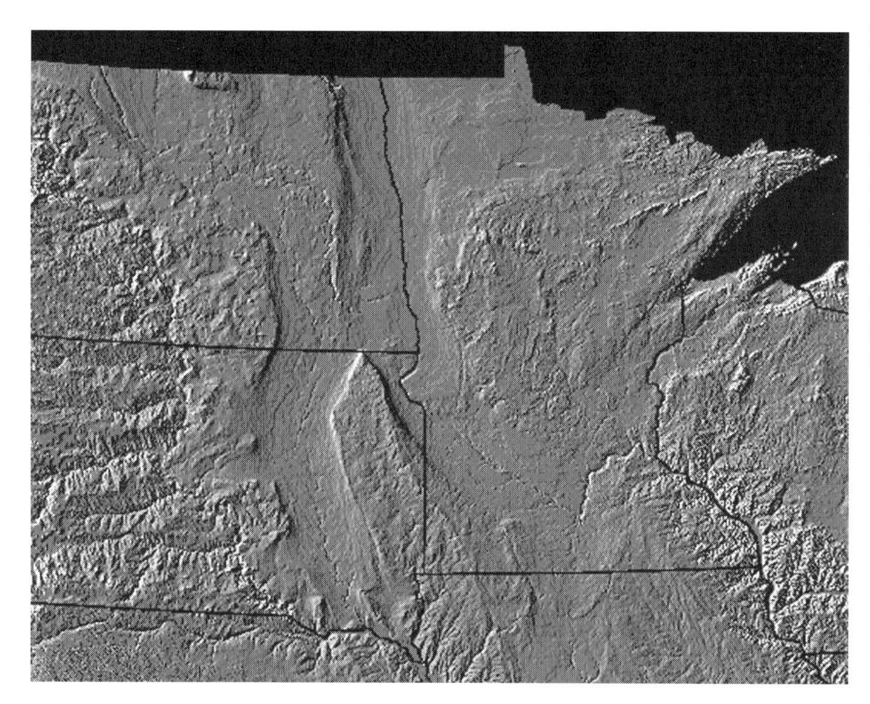

Figure 1.18 The landscape of the north central United States is shown as a simulated image. The effect of the southward moving great ice sheets on the terrain of eastern North and South Dakota and western Minnesota is evident by the broad channels that were eroded. Rivers now occupy the centerlines of these great valleys.
Credit: USGS.

sheets is obstructed by mountain ranges high enough to rise above the ice, the deepest, oldest ice is forced to flow uphill along the upstream sides of the mountains where wind and solar radiation ablate the surface. Such areas serve to conveniently concentrate diffuse meteorite fallout into rare "blue ice" areas where scientists can easily collect them. The name "blue ice" refers to the color of the bubble-free ice present at the surface of these regions. As the ice is removed, all the solid debris, primarily meteorites, contained in the ice and transported by the ice to this location, collects on the surface. These deposits are comprised of all material that fell on the ice sheet upstream of the blue ice areas.

Blue ice areas are sampling paradises where nearly all the rocks are of extraterrestrial origin. Dark meteorites are easily spotted against the bright ice background. There are only a few terrestrial rocks to compete for the attention of the meteoriticist because the interior of the ice sheet is far removed from sources of terrestrial debris. Over the past few years these areas have yielded more than 15,000 samples, many times the volume collected prior to the discovery of these treasured blue ice areas. Many of the samples have been grouped together into shower events; others are from exotic origins such as Mars, the Moon, and the asteroid Vesta. A precious few are completely unique specimens.

Summary

Ice has been, and will continue to be an important aspect of the Earth's climate and our habitation of this planet. Much has been learned about the behavior of this material and its uses by, and impact on, humans. It is more than metaphorically true to say that the future of research of ice is bright. In less than a century, we have progressed from a position, where attempts to travel across the ice sheets of the Arctic and Antarctic pushed the hardiest explorers to the limits of their abilities to survive, to the present, where we have established permanent bases where scientists are active year round in the pursuit of a wide range of knowledge. Whether from field data collected from these bases, or from data collected by satellites, scientists are using the ice sheets to help unravel the complex mechanisms that drive our planet's climate. Our terrestrially built knowledge base of ice, the meaning of its features, and the behavior of its many forms will be the basis for our interpretation of ice as it occurs and is explored on other bodies in our solar system and beyond.

BRYAN BUTLER
National Radio Astronomy Observatory

Ice on Mercury and the Moon

Is it possible that water ice could exist in the polar regions of Mercury and the Moon? At first glance it seems unlikely. These two bodies have no appreciable atmosphere and are surely bone dry in equatorial regions. On closer inspection, however, the idea is not so far-fetched. The question had been debated vigorously since the early 1960s in discussion of the Moon, but was never considered very seriously for Mercury.

In the summer of 1991, new radar maps of the surface of Mercury revealed features in the polar regions that seemed to indicate the presence of large amounts of water ice. Then, in 1998 and 1999, the Lunar Prospector spacecraft returned data that indicate an enhanced hydrogen content in the upper portion of the lunar polar soils, which some scientists have interpreted as evidence of water contained in lunar polar soil. How can this be? And if there is water ice at the lunar poles, how does that affect us?

Mercury

In spite of the fact that Mercury is one of our closest neighbors in the solar system, it is still one of the least probed and most poorly understood large bodies orbiting the Sun. Earth-based astronomical observations of Mercury are difficult due to the small separation of Mercury from the Sun in the sky (the maximum separation is only about 28 degrees). Before the Mariner 10 spacecraft arrived at Mercury in 1974 very little was known about the planet. We knew that the bulk density of the planet (about 5.5 grams per cubic centimeter) was very similar to that of the Earth and Venus, while its size (about 2,440 kilometers radius) is between that of the Moon and Mars. Its orbit is quite elliptical (e about .206), with an average distance from the Sun of about .387 AU, and an orbital inclination relative to the ecliptic of about 7 degrees.

Radar experiments had shown that Mercury is in an unusual orbital state, where for every three times it spins around its own rotational axis, it travels exactly two full circuits around the Sun. It takes the planet about 59 days to rotate once on its axis, and about 88 days to complete its orbit around the Sun. This makes the mercurian "day" equivalent to one full 3:2 cycle, or about 176 Earth days. Theoretical analysis of the orbit showed that in order for the 3:2 spin–orbit resonance to be stable, the rotational axis must be in one of two positions. Observations have since shown that Mercury's spin axis is nearly exactly perpendicular to the orbital plane. This will be an important point when discussing temperatures in polar regions later. Crude global optical albedo maps had also been constructed, but the quality was poor because of the difficult observing conditions. Clark Chapman, who made one of the most detailed of these pre-Mariner maps, later stated, "My own conclusion is that although it is probable that a few real albedo features were glimpsed occasionally, the errors in observation were so unavoidably large that, without the spacecraft, we would never have learned anything about Mercury's features."

Lessons from Mariner 10

During 1974 and 1975, the Mariner 10 spacecraft flew by Mercury three times, taking photographs of the sunlit portion of the surface each time, as well as gathering other valuable scientific data. Unfortunately, due to the particular combination of the spacecraft orbit with the orbit of Mercury, all three flybys occurred when the same part of the planet was illuminated, so only about half of the planet was photographed.

Half is better than none, however, and these photographs proved instrumental in shaping our ideas on the current state and history of the surface and interior of the planet. Other instruments returned equally valuable data that allowed a consistent picture of the interior and exterior of the planet to be constructed. The surface of Mercury appeared to be very similar to that of the Moon, with heavily cratered regions dominating the landscape (Fig. 2.1). There were certainly differences, like the conspicuous lack of "maria" (the dark lunar lava "seas") on Mercury, but for the most part the surfaces of the two bodies looked very similar.

A very thin atmosphere consisting of oxygen, helium, and hydrogen was detected. This atmosphere is so tenuous (many, many orders of magnitude less dense than that of the Earth) that molecules actually strike the surface of the planet more often than they encounter other molecules. Technically, this is considered an "exosphere" rather than an "atmosphere." We also learned that Mercury possesses a relatively

Figure 2.1 Mercury seen from the Mariner 10 spacecraft in 1974.
Credit: original images courtesy of NASA. Photomosaic of individual images courtesy of Mark Robinson, Northwestern University.

strong magnetic field, weaker than the Earth's but similar to it in structure.

Continuing to unravel the mysteries of Mercury

No spacecraft has visited Mercury since Mariner 10, and, as a result, progress beyond the knowledge gained from the Mariner 10 data has been very slow. One important new piece of information has been the discovery of relatively large quantities of sodium and potassium in the atmosphere of Mercury. The sodium and potassium are quite variable, and scientists have not reached consensus on the cause of the variability. Ironically, the lack of follow-up missions to study Mercury was generated, in part, by the Mariner 10 photographs. Because of the remarkable similarity of the surface of Mercury to that of the Moon, there has been a tendency to believe that nothing of great importance could be learned from Mercury that could not be learned from the Moon. We have recently discovered that Mercury is full of surprises, however, and still has much to teach us about the nature and history of the inner solar system.

It is important to understand Mercury more fully for several reasons (aside from sheer intellectual curiosity). Mercury actually has regions on its surface that are thermally acceptable for human habitation, although the radiation environment is particularly harsh.

Could we ever inhabit Mercury? If so, are there natural resources present that would aid human visitors or settlers? Data on Mercury may also place constraints on the collisional environment of the early inner solar system and solar system formation scenarios. If we can identify the materials that condensed out of the solar nebula to form Mercury, we may develop a better understanding of how our solar system formed and how solar systems are formed in general. The polar deposits will yield valuable clues about the volatile inventory in the inner solar system, possibly including mechanisms for delivery of water ice to the terrestrial planets. Mercury is also a great natural laboratory for the study of the generation of planetary magnetic fields. Finally, understanding the inner solar system is very important for adding to the body of knowledge about planet formation close to the center of solar systems, since many of the Extrasolar Giant Planets discovered in recent years are located relatively close to the "Sun" in their distant solar systems.

Mercury's proximity to the Sun, the 3:2 spin–orbit resonance, and the high ellipticity of the orbit cause surface temperatures in some equatorial regions to reach almost 800 K (1,000 F) during daytime, and plummet as low as 90 K (−300 F) during the long mercurian night. No other body in the solar system experiences such widespread extremes of temperature. If Mercury had a smooth surface, all of the surface would be sunlit during some part of the extended mercurian day, and it would be hard to imagine a place on the planet where it could remain cold enough at all times for water ice to exist. However, radar experiments in the early 1990s showed features near the poles of Mercury that are consistent with the existence of water ice in those areas. How can this be possible? Let us first examine the history of the discovery of these odd features, then consider what they might be and how they were emplaced.

Radar investigations

Because Mercury is particularly difficult to observe at optical and infrared wavelengths, longer wavelengths have been used successfully since the mid 1960s to probe the planet. Radar observations have been particularly useful, and, indeed, the discovery of the 3:2 spin–orbit resonance came from a radar experiment. Recent advances in two radar techniques have made information gained from radar experiments much more useful.

The first technique is the combination of two telescopes, one to transmit and one to receive, to make a combined true imaging radar system. The transmitting antenna is the 70 meter antenna at the Deep Space Network (DSN) complex in Goldstone, CA (operated by the Jet Propulsion Laboratory) (Fig. 2.2). This antenna transmits nearly 500 kilowatts of power, making it a very powerful transmitting telescope. The receiving telescope is the 27 antennas of the Very Large Array (VLA) in New Mexico (operated by the National Radio Astronomy Observatory). The signals received by the 27 VLA antennas are combined to make a "picture" of the radar reflectivity of the surface of the illuminated planet. This combined transmitting/receiving radar system was made possible in 1988 when the VLA antennas were outfitted with X-band (near 3.5 centimeter wavelength) receivers by NASA, in order to receive signals from the Voyager spacecraft as it passed by Neptune. Since

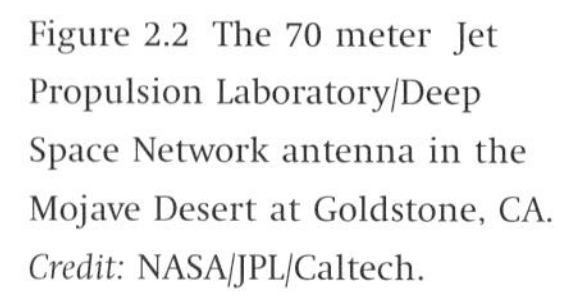

Figure 2.2 The 70 meter Jet Propulsion Laboratory/Deep Space Network antenna in the Mojave Desert at Goldstone, CA. *Credit:* NASA/JPL/Caltech.

Goldstone could transmit at this wavelength, it could be used to transmit a signal to a planetary surface, where it would be reflected and subsequently received and imaged at the VLA (Fig. 2.3).

The other technique was an enhancement to the traditional range-doppler radar technique that made mapping to the limbs of planetary surfaces possible. This technique, called the Coded Long Pulse (CLP) or random long code technique, was developed mostly at the Arecibo radar facility (Fig. 2.4). The Arecibo radar operates mainly at S-band (near 12.5 centimeter wavelength), and prior to 1998 operated with a transmitted power of about 420 kilowatts (this was later upgraded to nearly 1 megawatt). This technique is closely related to that used by Earth-based Side Aperture Radar (SAR) to map the radar reflectivity of the surface of the Earth, and by the Magellan spacecraft to map the radar

Figure 2.3 The Very Large Array (VLA) on the plains of San Augustin, New Mexico. *Credit:* NRAO/AUI/NSF.

Figure 2.4 The Arecibo Observatory in Puerto Rico. *Credit:* Arecibo/NAIC/NSF.

reflectivity of the surface of Venus. It is, however, fundamentally different from the Goldstone/VLA imaging technique (although they are sensitive to and measure a similar quantity). In both of these techniques, the transmitted signal is completely circularly polarized, with known circularity, and the full polarization state of the received signal is

Figure 2.5 Radar image of Mercury from a joint Goldstone/VLA experiment in 1991. The north polar feature is clearly seen at the top. *Credit:* D. O. Muhleman, B. J. Butler, M. A. Slade.

measured. This is done because, during the reflection of the signal from the planetary surface and subsurface, polarization changes occur, and these changes can be used to determine the properties of the surface and subsurface.

In 1991, both of these techniques were used for the first time to probe the surface of Mercury, by two different groups. The results were quite unexpected, and puzzling. In both experiments, the regions of the surface that had the highest radar albedo (were the best radar reflectors) were very near the north and south poles (Fig. 2.5). These polar features also had the very unusual characteristic that more signal was received in the same polarization as the transmitted signal than in the opposite sense. This was an unexpected result. Radar astronomers call it a "polarization inversion," since single reflections of the radar will result in a polarization reversal, and so more signal is generally received in the opposite sense to that transmitted. This peculiar radar signature is generally only observed in regions on solar system bodies that are known to have some kind of ice present, like the large moons of Jupiter, the polar caps of Mars, and portions of glaciers or ice and snow fields in Greenland and the Himalayas on Earth.

The currently accepted theory for the phenomenon of high radar reflectivity accompanied by a polarization inversion is that of "coherent backscatter." This phenomenon is also sometimes referred to as "weak localization" or "time reversal symmetry." The physical model for this effect has the planetary area being probed by the radar made up of a material that is mostly transparent to the radar waves. A good portion of the energy in the radar wave therefore penetrates through the upper surface of the material. However, embedded in the material are many "scattering centers" (cracks, voids, or buried inclusions, for example) that are capable of interacting with the radar wave and changing its direction and polarization properties. So, after penetrating the material, the radar photons bounce around from scattering center to scattering center, changing their characteristics (energy and polarization) at each bounce. Some of the photons escape from the surface in exactly the right direction to be captured by the receiving radio telescopes.

Detailed treatment of the physics of the scattering problem for the geometry of this type of physical structure has shown that both high radar albedo and a polarization inversion can result if the scattering centers have the right properties. Because water ice (and particularly cold water ice) is very transparent at microwave frequencies, it is a candidate for the material containing the scattering centers. The scattering centers could be cracks or voids in the ice, which would be very long-lived, especially in cold ice. Ices of other molecules may also be very transparent to microwaves, but we do not know for sure due to a lack of data regarding their microwave properties.

What are these features?

From the initial Arecibo observations, it was clear that the feature near the south pole was related to the large (150 kilometer diameter) crater Chao Meng-Fu. The feature near the north pole seemed to be composed of several regions, and was somewhat more confusing and

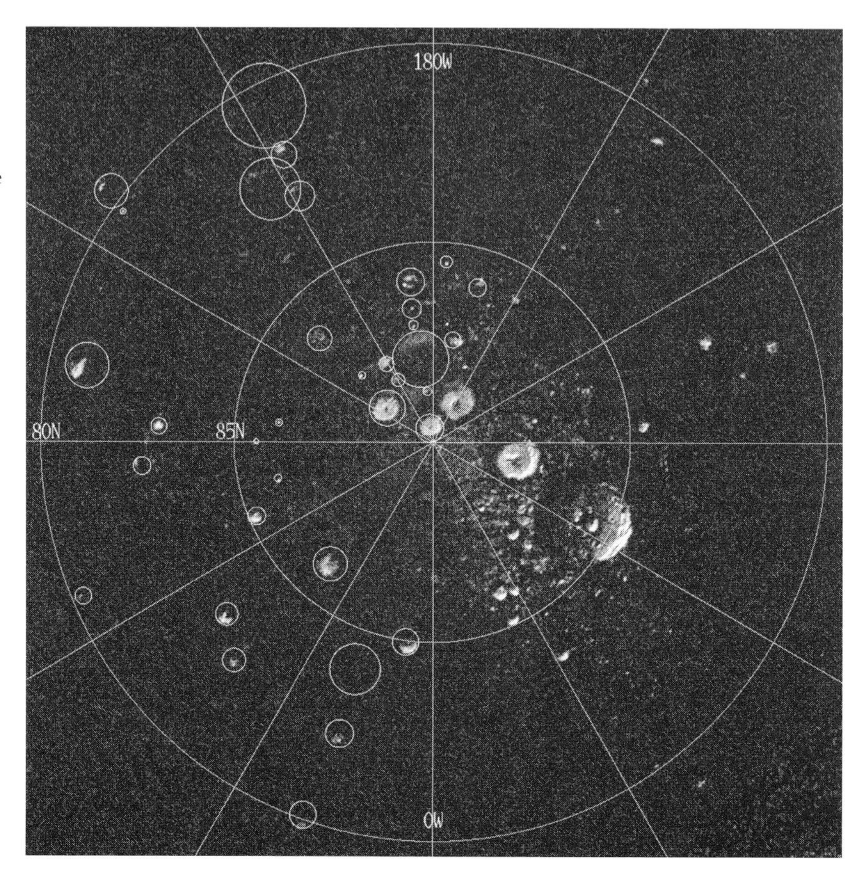

Figure 2.6 Radar image of the north polar region of Mercury from Arecibo observations in 1999 with 1.5 kilometer resolution. Light areas denote higher reflectivity. Circles denote the locations of craters as imaged by the Mariner 10 spacecraft.
Credit: J. K. Harmon/NAIC.

harder to interpret with the initial data. Further experiments from Arecibo (after the power upgrade) produced extremely good, high-resolution (1.5–3 kilometers per pixel) images of the north polar region. Unfortunately, the south polar region has not been accessible to Arecibo since the upgrade, and so the images of that region are not quite so good (resolution only about 15 kilometers), but are still quite spectacular. Where these radar images could be compared with the photographs from Mariner 10, it was immediately apparent (even from the lower resolution images) that the high albedo, polarization inversion regions were all located within the boundaries of craters (Fig. 2.6). The puzzle was then to figure out what could be causing these features, and, specifically, what is in the polar craters of Mercury that causes the very high radar reflectivities and polarization inversions observed there?

There have, to date, been five proposed explanations for the measured radar signatures. First, the features could be caused by very fresh lava flows. Fresh lava flows on Earth, Mars, and Venus have been observed to have very high radar albedoes and have polarization signatures that are similar to the Mercury polar features (but not quite so strong). The argument against lava flows is geological. Why would there only be flows in polar craters? Also, the photographed portion of Mercury's surface shows no recent geological activity, so fresh lava flows would be unlikely.

Second, it could be that the amount of free sodium and potassium in the very thin atmosphere of Mercury could cause collections in polar craters, which may cause the rocks there to alter into something which could give the required radar signature. Polar enhancements of atmospheric sodium have been observed on Mercury. The argument against this hypothesis has been that, while this might give the required radar albedo, the polarization signature would probably be wrong.

Third, the signature could be caused by the incoming radar wave interacting with very cold rocks. It is clear that the coldest places on the planet are the floors of polar craters, so the rocks there will be as cold as anything can be on Mercury. The argument against this is the same as the argument against the sodium/potassium explanation. Furthermore, the cold polar craters of the Moon show no similar radar signature.

Fourth, it may be possible that very cold rocks fracture (from incoming meteorite bombardment) in a way that encourages very angular, jumbled rock formations. These formations might have radar characteristics similar to the very young lava flows. However, while such formations may cause slight polarization inversions, they do not cause inversions of the strength observed in the polar crater deposits, and, furthermore, the lack of these radar signatures in lunar polar craters also argues against this hypothesis.

Lastly, there could be large deposits of volatile materials (ices) in mercurian polar craters. This last explanation is the most likely and currently the most accepted of the five. This is because this type of scenario explains all of the observed data. Some types of volatile deposits do have very high radar albedoes and inverted polarization signatures, and volatile deposits would be most likely to form in the cold polar craters.

Volatiles on Mercury

What is meant by "volatile materials" here? A "volatile" is any molecule that can turn into vapor at the existing temperature and pressure of its environment. Normally, volatiles are considered to be those molecules that are readily vaporizable at low temperatures: water, carbon monoxide, carbon dioxide, sulfur dioxide, ammonia, etc. In their solid form, these volatiles are called "ices." On Mercury, however, "ice" takes on a somewhat unusual definition. Because of the extremely high temperatures on Mercury, more stable molecules may actually be volatile. Therefore, the possibility exists that there could be deposits of materials which are normally not considered as volatiles, that could still be formally called "ice" deposits.

Sulfur has been suggested to explain the radar features on Mercury. Other examples are the sodium and potassium (if, for example, instead of interacting with the rocks to alter them the sodium or potassium simply collected on top of the rocks to form thick deposits). There are several arguments against these "warm" volatiles as the primary constituent of the deposits observed in the radar images. One is that their radar properties do not in general satisfy the criteria necessary to create the two observable radar signatures: high reflectivity and polarization inversion. Additionally, there are problems accumulating enough material to form the observed deposits. And, lastly, they are too stable. This will be discussed in more detail later.

How much volatile material is needed to form the deposits? In order to determine this quantity, an estimate of the total area covered by the deposits is needed. In addition, some estimate of the depth of the deposits is required. From the best Arecibo radar imagery, the total area covered by the deposits is about 30–50 thousand square kilometers (roughly the size of The Netherlands). This is less than 0.07% of the total surface area of Mercury. The depth is somewhat harder to estimate, since it is linked to the detailed physics of the radar wave scattering in the deposits. The best estimates suggest a depth of at least 5–10 meters. This means that a total volume of volatile material of order a few hundred cubic kilometers would be sufficient to create the deposits (somewhat less than the volume of Lake Erie).

Necessary conditions for ice at Mercury's poles

How, then, can ices exist in the polar craters of Mercury, a planet so near the Sun? Three conditions must be met for ice deposits to form and persist over time:

1. there must be delivery of a sufficient quantity of the volatile material to the surface,
2. that volatile material must be able to migrate to regions where it is stable (the polar craters) and a sufficient quantity of the volatile must survive that migration, and
3. the environment in these locally stable locations must allow the condensed volatiles to remain stable for long periods of time (otherwise we must presume that we are looking at Mercury at a special time).

Delivery of volatile materials to the surface of Mercury could be either endogenic (from the planet itself, from volcanic outgassing or direct degassing from subsurface rocks), or exogenic (from outside the planet, from meteoritic, cometary, or solar wind material). Mercury itself is thought to be relatively volatile poor, because of the spectral signature of its rocks and soil, and the extreme age indicated by its heavily cratered surface. Although it seems unlikely that the source of the volatiles is endogenic, an endogenic source cannot be ruled out entirely at this point. There are several possible exogenic sources of volatiles to the surface of Mercury, including: the constant micrometeorite bombardment of the surface, larger meteorites (up to asteroidal size), and comets (long and short period).

Once a volatile molecule is delivered to the surface, it migrates through a number of ballistic hops until it is either photomodified (broken apart or ionized by solar radiation) or lands in a stable region (a region where it is no longer volatile and condenses). The stable regions are those that are the coldest on the planet – near the poles and on the nightside. However, if a molecule makes it to the nightside, its stability is short lived, as that location will be brought into sunlight as the planet spins, and the molecule will begin its journey anew. Polar regions are, therefore, the only regions where long-term stability is possible.

After formation of a volatile deposit in to a stable polar region, there are several processes that can destroy that deposit. The most important of these processes are: thermal evaporation (sublimation), destruction by energetic particle sputtering or solar radiation, and erosion by micrometeorites. In fact, it has been shown that the rate of loss of any reasonable volatile from all processes other than thermal evaporation would be so great that it is unlikely that any of these processes are in operation. The only way to avoid these processes is if the volatile deposits are covered by a relatively thin layer of regolith. It must be thin (less than roughly 10 centimeters) or the radar would not probe through it. This thin layer would protect the underlying volatile deposit from micrometeorite and energetic particle erosion, and from the intense scattered solar radiation (mostly Lyman-alpha) that could destroy the deposit. Even the rate of thermal evaporation by itself may be too great to allow the long-term survival of the deposits if the temperature of the deposits is not cold enough.

Constraints from supply and temperature

How cold is it in the craters near the poles of Mercury? The temperature of interest is the maximum temperature throughout the extent of the entire mercurian day (176 Earth days). This is because the total amount of evaporation is determined almost entirely by the evaporation during the time when the deposits are at their warmest. This maximum temperature for any location on the planet is determined by the maximum amount of radiation incident upon that location. The radiation can issue directly from the Sun, can be solar radiation reflected from a nearby surface (like a crater wall), or thermal radiation from below (subsurface heating).

The maximum amount of direct solar radiation into polar regions is determined by the obliquity of the planet's rotation axis. For example, the Earth's obliquity is near 23.5 degrees, so at one time during the year, the Sun is approximately 23.5 degrees above the horizon at either pole. Theory and observation have both shown that the obliquity of Mercury is near zero degrees (the rotation axis of the planet is nearly exactly perpendicular to the orbital plane) so the amount of direct solar radiation reaching the polar regions of Mercury is quite small.

In fact, direct radiation only reaches the exact pole because the Sun appears very large in the sky (about 1.5 degrees apparent diameter), due to its size and relative proximity to Mercury. So, at the poles, the limb of the Sun peeks over the horizon at all times. Even this small amount of high-angle solar radiation is enough to warm the poles up to temperatures near 175 K during the warmest part of the mercurian day. However, there can be large areas that never see any direct sunlight in the floors of polar craters. They are always shaded on their equatorward sides by the crater walls. In these permanently shaded regions, detailed thermal models show that temperatures can remain very cold throughout the entire mercurian day. The exact maximum temperature is determined by the precise shape of each crater and its latitude, but typical maximum temperatures in the permanently shaded parts of the larger polar craters are less than 100 K, and they can be as cold as 60 K.

With combined information on supply, migration, and stability of a number of volatiles, as well as information on the observed deposits and estimated temperatures, it is possible to estimate the probability that the deposits are composed of a particular volatile, and hence identify the most likely volatile material.

The first consideration is that the observed deposits are only in the floors of craters, not in true polar "caps." This means that volatiles that evaporate very slowly at temperatures above about 200 K are unlikely to be the major constituent of the deposits. For example, it would take a 1 meter deposit of sulfur about a billion years to evaporate at a temperature of 225 K. This means that if the observed deposits were made up primarily of sulfur, then polar "caps" should also be observed, which is not the case.

The other "warm" volatiles are similar – they are just too stable to be the main ingredient of the deposits. On the other hand, volatiles that are not stable enough to last for long periods of time at temperatures above about 100 K are also unlikely to be the major constituent of the deposits. For example, a carbon dioxide ice deposit would have to be colder than about 60 K to last for a billion years. Other ices like ammonia and sulfur dioxide are similarly unstable. There are some regions in some mercurian craters for which the maximum temperature is this low, but the total area covered by these regions is very small. This means that there could be very small amounts of these more unstable ices in isolated regions, but they do not explain the majority of the observed radar features.

Some ices have the right temperature characteristics to make them likely candidates for the majority of the deposits (they would not form "caps" at temperatures around 200 K, but are stable for long periods at temperatures around 100 K), but their supply is too small. Examples of such ices are HCOOH (formic acid) and CH_3OH (methanol).

Water ice – just right

Water ice, however, has the right temperature characteristics to make it a likely candidate for the majority of the deposits, and water ice is sufficiently abundant to supply the required amount of material. At a temperature of 150 K, it takes only about a thousand years to evaporate a 1 meter thick deposit of water ice. So, true polar "caps" would not last for very long, explaining their absence in the radar images. Lower the temperature to just 110 K, however, and that same 1 meter thick deposit of water ice will take almost a billion years to completely evaporate. Detailed analysis has shown that about 10% of water molecules migrating around the surface will reach the polar cold traps. Further analysis has shown that impacts of short period Mercury-crossing comets (either extinct or active) or larger meteorites could supply the required amount of water to the planet.

A likely scenario then emerges for the creation of the polar ice deposits on Mercury that involves three major elements. First, one (or several) large short period Mercury-crossing comets (possibly even extinct) or meteorites (large enough to be considered asteroids) smacked into the surface, delivering a large amount of water to the planet. The water molecules left over after the collision then migrated to permanently shaded regions in polar craters. The water ice deposits were finally covered up by regolith "fallout" from other nearby crater-forming impacts.

This interpretation is compelling, but it is not conclusive. It is not absolutely certain that the observed radar features are even ice deposits of any kind, much less water ice deposits. This is because there is currently no single piece of evidence available to determine what is causing the observed signal in the polar craters. There are, however, many measurements that could be made either to determine directly what causes the features, or to provide valuable information to help in that determination. Note that since these deposits are thought to exist in permanently shaded regions, they will probably never show up in a normal photograph, since there is no incident sunlight to be reflected back to the camera. Therefore, good photographs of the polar regions may not help in this respect (it would, however, be very nice to have photos of the unphotographed side of Mercury in order to determine if the craters on that side exhibit similar features to those identified on the side photographed by Mariner 10).

Precise determination of the surface and near-surface temperatures in and around the polar craters on Mercury would help to constrain the types of volatiles that could remain stable in the polar regions for long periods. This could be done with a spacecraft thermal mapper (thermal IR or longer wavelengths), or with very high-resolution ground-based measurements. More precise mapping of the deposit locations would provide important information on spatial morphology. When combined with temperature measurements, or with accurate thermal models, this could also help to constrain which (if any) of the volatiles could be responsible.

This could be done with a spacecraft radar mapper (like Magellan). Searches for molecular emission above the poles may detect the vapor of the volatile (or one of its byproducts) directly. For example, hydroxyl (OH), sulfur compounds, or carbon monoxide (CO) might be detected. These molecules (among others) have strong emissions (UV, IR, or submm/mm) that could be detected with a spacecraft, an earth-orbiting satellite, or possibly even from the ground.

Finally, measurement of the spectrum of neutrons above the poles may provide indirect evidence that the deposits are water ice. Hydrogen has a distinct neutron signature that could be distinguished with a neutron spectrometer aboard a spacecraft (see the discussion in the "Moon" section below). Two spacecraft missions to Mercury are in the planning stages, and these will both carry instruments that may help to answer some of these questions (see the "Future missions" section below).

Lessons from ice on Mercury

So, it seems likely that there are large deposits of ice (probably mostly water ice) in permanently shaded regions in polar craters on Mercury.

What can we learn from such deposits, or what good are they to us? Well, the fact that these ices even exist can place constraints on the current and past thermal environment of Mercury. Their existence may also constrain the past orbital history of the planet. Constraints on the past meteoritic and outgassing environment may also emerge if more information is gained about the true extent and amount of ice. Constraints on mechanisms for delivery of water to the surfaces of the terrestrial planets could be obtained.

If we could ever send a robotic or human mission to the polar regions, an ice core could be obtained and this would be extremely helpful in untangling the past history of Mercury. Just as Antarctic ice cores are used here on the Earth, such a core for Mercury could provide clues about any atmosphere that may have existed. Such a core could be used to determine whether the ice was delivered in one or a few events, or slowly over time (due to possible isotopic and contaminant differences with depth), yielding further clues to unraveling the surface bombardment history. It could also be used to precisely determine the composition of the ice. In the realm of wild speculation, if Mercury is ever inhabited by humans, these polar ice deposits would be an available natural resource.

The Moon

The Moon is at the other end of the spectrum from Mercury in terms of how much we know about large bodies in the solar system. We know more about the Moon than any other large body in the solar system besides the Earth (Fig. 2.7). It is the only body that astronauts have visited and it is the only body for which we are certain we have samples because they were collected by astronauts and spacecraft specifically for study in laboratories on Earth. We know the physical characteristics of the Moon (orbit, size, bulk density, etc.) very well. Yet there are some things that we still do not know about the Moon. We do not know the characteristics of the polar regions of the Moon, or specifically, whether water ice is present at the lunar poles.

What do the Mercury studies described above tell us about the possibility of ice in the polar regions of the Moon? By implication, if such ice deposits exist on Mercury, they may exist on the Moon. The existence of water ice in the polar regions of the Moon has been the subject of

Figure 2.7 Photograph of the Moon taken by Apollo 17 crewmen during their homeward journey in 1972. Mare seen in this photo include Serenitatis, Tranquillitatis, Nectaris, Foecunditatis, and Crisium. *Credit:* NASA.

long-standing and vigorous debate. The concept was proposed and discussed as long ago as the 1700s, but until the mid twentieth century was more in the realm of romantic conjecture. The modern discussion of possible ice on the Moon began in the 1950s and 1960s. Researchers at that time concluded that ices, and particularly water ice, could indeed exist in permanently shaded regions near the poles of the Moon.

Discussion of this idea lagged somewhat during the Apollo years, and one piece of data from those missions seemed to indicate that there might be no polar ice on the Moon. The lunar soil and rock samples seemed to be incredibly desiccated. There were literally no detectable water-bearing minerals in the lunar rocks and soil returned for examination. Meteorites thought to come from the Moon show a similar lack of water-bearing minerals. This seemed to imply that the Moon was indeed a very dry place, and cast a shadow of doubt on the idea of water ice there, even in permanently shaded polar regions.

Re-examining the case for water ice on the Moon

Detailed re-examination of the idea in the late 1970s, prompted by the Apollo findings, indicated that it was still possible that water ice existed in the permanently shaded regions near the poles. In the years following the race to the Moon much effort has been devoted to trying to show that no water ice should exist in the polar regions, for any of several reasons, including: possibly inefficient transport to the polar regions, erosion of deposits by energetic particles, and erosion by radiation from the local interstellar medium.

More recent modeling has shown that transport should be sufficiently efficient to form the deposits (much more efficient than for Mercury, in fact), and erosion of the deposits can be avoided by covering the ice with a thin layer of soil (similar to the Mercury case discussed above). Temperatures in the permanently shaded portions of lunar polar craters have also been shown to be as cold as on Mercury, so the stability of the ice should not be a problem.

So, if enough water has been delivered to the surface of the Moon, it is certainly plausible that water ice deposits exist in the permanently shaded craters in the polar regions of the Moon. Since the identification of the probable water ice deposits in the polar regions of Mercury, several experiments have been performed in an attempt to find such deposits on the Moon.

Radar investigations

The first new piece of evidence was obtained from high-resolution radar images of the lunar polar regions obtained in May and August of 1992. The Arecibo telescope was again used to obtain images, and radar maps of the two polar regions were produced, with a resolution of about 125 meters (Fig. 2.8). These maps were examined closely for signatures similar to those seen in the Mercury polar regions. No such signatures were found, implying that there are no large (greater than about 200 meters diameter) ice deposits in the polar craters of the Moon that were probed by the radar.

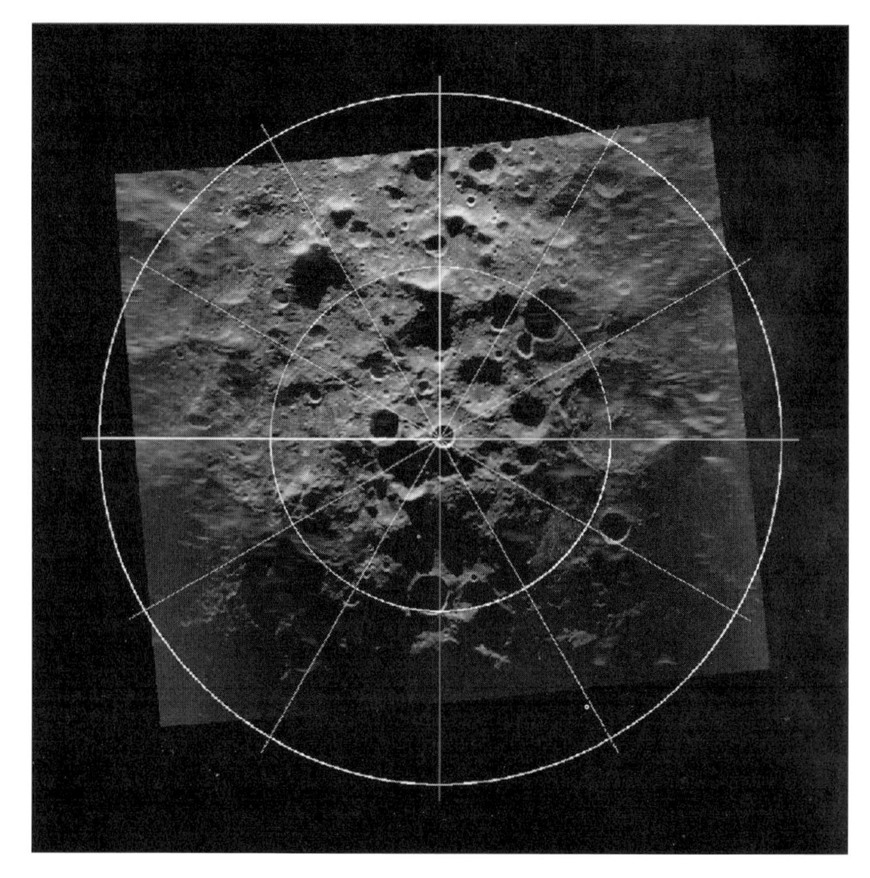

Figure 2.8 Radar image of the south polar region of the Moon from Arecibo observations in 1992. Resolution approximately 500 meters. Lighter areas denote higher reflectivity. Latitude circles are at 85 and 80 degrees south.
Credit: N. J. Stacy.

One other result from this experiment suggested that it was much more likely that any deposits, if they existed, would be located in the south polar regions rather than the north polar regions because the north polar regions are much flatter than the south polar regions. As a result, more permanently shaded areas exist in the south polar region than in the north.

Clementine and the hunt for ice at the lunar poles

The relative amount of permanently shadowed terrain in the two polar regions was verified by scientists examining images from the Clementine spacecraft. By combining all of the photographs of the north and south polar regions, it was possible to determine which regions were never illuminated during the periods when the photos were taken. It was clear from these photos that many more locations (covering a much larger accumulated area) in the south polar region remain in perpetual shadow than in the north polar region. Later radar experiments designed specifically to determine the topography in the polar regions confirmed this result, showing that there is about twice as much permanently shadowed terrain in the south polar regions (roughly 5,000 square kilometers) as in the north (roughly 2,500 square kilometers).

In April of 1994, the Clementine spacecraft attempted to detect polar water ice deposits directly. During seven orbits, the transmitting antenna on the spacecraft was directed toward the polar regions, and the resulting reflected signal was received at one of the 70 meter DSN antennas here on Earth. During one of the orbits that passed over the south pole, the data showed a signature that is somewhat similar to the polarization inversion seen in the Mercury radar data. This was interpreted as the signature of a small deposit of water ice about 100 square kilometers in area, or only about one-thirtieth the size of Rhode Island. This is much smaller than the size of the proposed deposits on Mercury. Subsequent detailed re-analyses of those data have either supported the water ice argument (in fact even locating it precisely in Shackleton crater), or shown that there is not only no water ice signature, but no signature at all in the data. There is still no wide agreement regarding the signature of ice in the Clementine radar data.

Lunar Prospector

At approximately the same time, the Lunar Prospector mission was chosen as the first competitively selected NASA "Discovery" mission. Prospector was launched on January 6, 1998. The trip to the Moon is short, and on January 11, the spacecraft was successfully maneuvered into its polar orbit. This spacecraft was especially important with respect to the question of ice in the polar regions of the Moon, since it was equipped with instruments that would return data to address the issue of ice at the lunar poles. The spacecraft carried neutron and gamma ray spectrometers, which are sensitive to very energetic particles, and could be used to measure the spectrum of neutrons coming off the lunar surface.

As mentioned above, the spectrum of neutrons observed coming off regolith which is rich in water ice is distinctly different from the spectrum that is observed from normal regolith. How does this work? When a cosmic ray strikes a planetary surface, it has so much energy that it penetrates deeply into the surface. Along its path, it smashes into atoms, releasing their neutrons into the surrounding material. These neutrons bounce around in the material, interacting with other atoms in it. During each interaction with an atom, some amount of energy is lost from the neutron. If the atom is a large one, then the neutron only loses a small amount of energy. However, if the atom is a small one (like hydrogen), then a much larger amount of energy is lost. Surfaces that have an overabundance of hydrogen atoms will have an excess of neutrons coming out, with energy determined by the temperature of the surface ("thermal" neutrons), and a deficiency of neutrons with more energy ("fast" neutrons), or mid-range energy ("epithermal" neutrons). If it is possible to measure the energy spectrum of neutrons coming off a planetary surface, then it can be inferred that the surface has an overabundance of hydrogen atoms. Since water is such a common solar system volatile, it can then be concluded that an overabundance of hydrogen indicates the presence of water molecules.

Lunar Prospector neutron spectrometer data

Very early in the Prospector mission, it was announced that a signature had indeed been seen in the neutron spectrometer data that indicated the presence of water ice in polar craters. Specifically, a decrease in the amount of epithermal neutrons was seen over both poles. Because of the small size of the decrease in the signal, it was inferred that the excess hydrogen being measured was mixed intimately with the surrounding regolith, at mixing ratios of about 1% (which implies a water mixing ratio of about 10% if all of the excess hydrogen is contained in water molecules).

The total volume of water implied by the decrease is about 10 cubic kilometers at the north pole, and about half that amount at the south pole. This assumes that the water ice is mixed in with the regolith in the upper 0.5 meters (about the maximum depth to which the instruments sample). Later in the mission, the orbit of the spacecraft was lowered to within a mere 10 kilometers of the surface. As a result, the data gathered after that time had much better resolution (about 15 kilometers per pixel). Detailed analysis of the higher-resolution data shows that indeed the increased hydrogen signature is mostly confined to the interiors of the polar craters, similar to the Mercury polar radar features.

However, it still has not been shown conclusively that the increased hydrogen signal is due to water ice. In fact, it has been argued that the decreased epithermal signature is not due to hydrogen at all, but rather to mineralogical differences in the polar soil and rocks. This is because, although a decrease in the epithermal signature was seen, no corresponding increase in the thermal signature was observed (or at least reported), as would be expected if there was an overabundance of hydrogen in the soil and rocks.

Furthermore, it has been argued that, even if there is such an overabundance, the increased hydrogen could simply be caused by solar wind protons. Since the lunar regolith is saturated with hydrogen, solar wind protons that impact the surface cannot be accommodated there. So, upon impact, various hydrogen-bearing molecules are created that can subsequently migrate to the polar regions. Water ice is only a small fraction of the resulting volatile deposit that is formed. This mechanism is not valid for Mercury because its relatively strong magnetic field protects the surface from the solar wind most of the time. This explanation is disputed by some, and there is no wide agreement on the issue.

Whether the excess hydrogen in the polar regions of the Moon is indeed water ice or not, several questions remain. First, if it is ice, why is the ice in the lunar polar regions so different from that which we think is present in the polar regions of Mercury? Remember that to be seen in the radar data, the ice deposits have to be very clean, pure, ice deposits, and must be quite thick (several meters). This is very different from the picture presented by the Lunar Prospector data, where the ice is present in smaller quantities, and likely mixed in with the regolith. One possible answer to this is that the Mercury ice may have been delivered in a small number of events (possibly from large short period comets), while the Moon ice was delivered slowly over time, through the constant bombardment of tiny micrometeorites, or solar wind protons which recombine to make water molecules. Another is that the small obliquity variations of the Moon cause the ice in polar craters to be somewhat less stable there than on Mercury.

Figure 2.9 Two jubilant scientists on an ice-prospecting lunar lander mission examine an ice-encrusted drill stem as they stand in the frigid (60 K) permanently shadowed part of the south polar region crater. *Credit:* original painting by Mike Stovall provided courtesy of NASA.

Secondly, why would there be more hydrogen (or possibly ice) in the north polar regions of the Moon than in the south polar regions? Both the Earth-based radar data and the Clementine photographs seem to indicate that there is much more permanently shaded terrain in the south than in the north. Is more hydrogen (or water) delivered to the northern hemisphere of the Moon, and, if so, why? Is this an outgassing signature? This very odd situation awaits explanation.

If it can be shown that these deposits are indeed water ice, then this is an incredibly important discovery. Even if they are not, that finding will be significant. These deposits would be of tremendous scientific value, as they would provide valuable information on past thermal, solar, and collisional conditions in the Earth–Moon system. They would also be a possible natural resource for future human missions and/or colonies on the Moon (Fig. 2.9). Water ice on the Moon could be used to sustain lunar colonies, both by providing the water needed for survival, and in providing the oxygen needed to breathe. Excess hydrogen in any form could also be used to provide rocket fuel needed for either a return to Earth, or for launching rockets toward other destinations in the solar system.

However, these possibilities involve both getting to the polar regions of the Moon cheaply, and, once there, getting the water or other hydrogen-bearing molecules out of the lunar rocks and soil efficiently.

Currently, there is no reliable and cheap method for extracting the water, but this is a very active area of research. Possibilities include simple heating of the rock/soil and electrolysis. It seems reasonable to assume that if these deposits are proven to exist, efficient methods for getting to the poles and for extracting the water will be developed.

Future missions

While it is possible to find out more about these deposits on both Mercury and the Moon via Earth-based observations, observations from spacecraft (and possibly landers and/or rovers) will really be needed to conclusively show the composition of the deposits. Three new lunar missions are on the horizon. The ESA (European Space Agency) Smart-1 mission launched on September 27, 2003 (but will take 2–2.5 years to reach the Moon because it is testing ionic propulsion); and two ISAS (Institute of Space and Astronautical Science of Japan) missions. Lunar-A is scheduled for launch in 2004 and Selene should launch in 2005 (a joint mission with NASDA–National Space Development Agency of Japan).

These three missions, while not explicitly designed to answer questions about the polar deposits, will certainly return data that will help to constrain the possibilities regarding the composition of the lunar polar deposits. For example, Lunar-A will carry heat probes that will help to measure the heat flow from the interior of the Moon, which is an important parameter in the models that determine the temperatures in the shaded polar craters.

Two missions to Mercury are planned: a NASA mission, MESSENGER, and an ESA mission, BepiColombo. These spacecraft will carry instruments designed specifically to address the nature of the polar deposits. Both will carry neutron and gamma-ray spectrometers similar to the Lunar Prospector spectrometers. Both spacecraft will carry imaging cameras, which will photograph the hemisphere not illuminated during the Mariner 10 flybys, thereby helping to determine if polar craters on the side of Mercury that has never previously been photographed exhibit features similar to those observed in polar craters photographed by Mariner 10.

MESSENGER will carry a spectrometer capable of identifying byproducts of the polar deposits via their ultraviolet or infrared emission signatures in the tenuous atmosphere of the planet. MESSENGER will also carry a laser altimeter that will help constrain topography in the polar regions (although it will not explicitly measure the topography above 80 degrees latitude).

BepiColombo will have a surface lander that may land at high latitude. If so, this would help to determine the environment at such high latitudes on Mercury. MESSENGER is scheduled for launch in the summer of 2004. The spacecraft will go into orbit around Mercury in 2011. BepiColombo is somewhat later, with a planned launch around 2010.

The data returned from these spacecraft, along with Earth-based observations, will help us to narrow down the possibilities for the composition of the polar deposits on both the Moon and Mercury. It is to be hoped that further in the future, landers, probes, and rovers will be sent to the polar regions of both Mercury and the Moon to truly determine the composition of the deposits. If it turns out that the deposits are indeed water ice, what a wonderful discovery!

TOBIAS OWEN
Institute for Astronomy, University of Hawaii

How the Earth got its atmosphere

Why does the Earth have air and oceans? Mercury and the Moon have essentially no atmospheres at all, while Jupiter and the other giant planets have atmospheres more massive than the entire Earth. Our own atmosphere is mostly nitrogen, whereas our nearest neighbors Mars and Venus both have atmospheres dominated by carbon dioxide.

The thin atmosphere of Mars produces a surface pressure that is only 1/150 the sea level pressure on Earth, while the atmospheric pressure on Venus is 90 times the terrestrial value. In the outer solar system, hydrogen and helium are the most abundant gases, except on Triton, Titan, and Pluto, where once again nitrogen rules the roost. But these are very different roosts (Fig. 3.1). What accounts for all these differences? Where do all these gases come from? What does any of this have to do with ice?

How planets keep their atmospheres

All planets are surrounded by the vast, uncompromising emptiness of space, which even in the solar system contains fewer atoms in each cubic centimeter than the best vacuums we can produce in our laboratories. So we might begin our inquiry by asking how planets can have atmospheres at all. Why don't the atoms and molecules making up these gases simply fly off into space?

The ability to maintain an atmosphere over the lifetime of the solar system depends on a planet's gravitational field, the composition of the atmosphere and the temperature of the "exosphere" – the outermost layer of the atmosphere from which gases can escape into space. The density of our own atmosphere steadily decreases with altitude until we reach a level about 500 kilometers above the ground, where a molecule heading upward will no longer encounter another molecule. This is the base of the exosphere.

All that is necessary for escape into space is that the kinetic energy (or energy of motion) of a moving particle, be it a gas molecule or a spaceship, is equal to its potential energy (the energy it owes to its position in the Earth's gravitational field). The lightest molecules move the

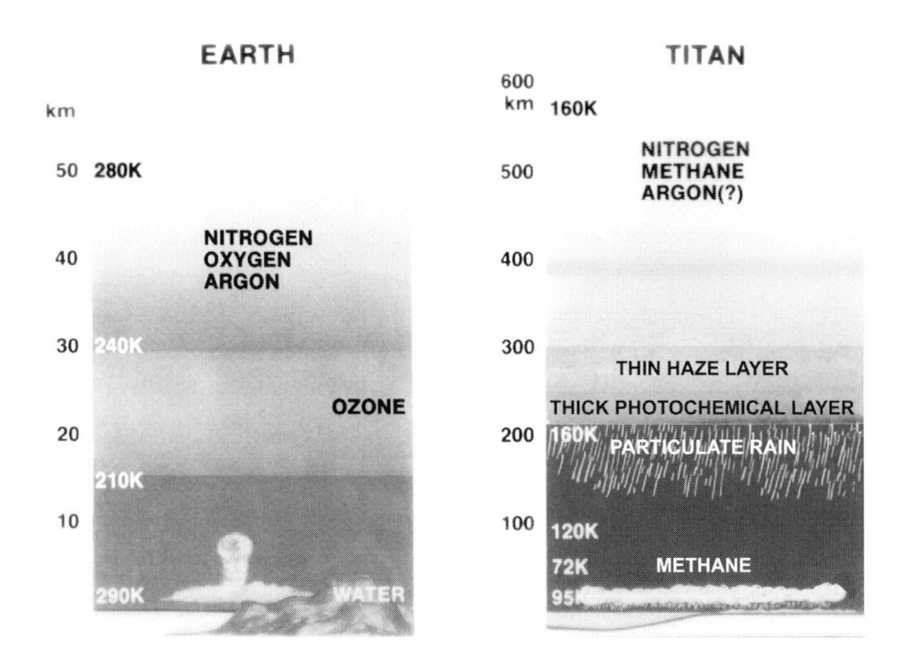

Figure 3.1 Saturn's satellite Titan is larger than the planet Mercury and has a nitrogen-rich atmosphere more massive than our own. But methane, rather than oxygen is the next most abundant constituent, owing to Titan's low surface temperature. Both methane and ethane may condense in the lower atmosphere, raining down on the surface to produce rivers and lakes on the icy landscape. Smog particles settling out from the photochemical haze will add to this flammable landscape. *Credit:* NASA.

fastest, and only the fastest molecules can escape. On our own planet, hydrogen and helium can flee into space with ease, while heavier gases such as neon, nitrogen, and oxygen remain in the atmosphere essentially undiminished over the lifetime of the solar system.

In other words, the Earth retains its atmosphere because its mass is sufficient to produce a gravitational field that prevents the atmospheric gases from escaping into space. Our Moon, with one-eightieth the Earth's mass, can only maintain an extremely tenuous, flickering envelope of atoms. As atoms and molecules escape, they are resupplied by gases leaking from the lunar interior and others carried to the Moon by the solar wind. Giant Jupiter has 318 times the mass of Earth and is therefore able to preserve an atmosphere made mainly of the lightest gases: hydrogen and helium. These are also the two most abundant elements in the universe, the major components of stars.

We thus say that Jupiter and its fellow giant planets have retained primitive atmospheres, close to the composition of the original solar nebula. With this huge excess of hydrogen, it is not surprising to find that these primitive atmospheres also contain many hydrogen-rich compounds, such as methane (CH_4), ammonia (NH_3), and water (H_2O). The less massive inner planets have long ago lost any original hydrogen and helium they once contained.

Why small planets have different atmospheres

So we now have one clue to explain the different compositions of the atmospheres of different planets: mass matters! Giant planets can retain even the lightest gases, while really small objects can not even hold on to gases like xenon, which has 65 times the mass of molecular hydrogen. We might expect Venus, Earth, and Mars to have very similar atmospheres, since they are all small rocky planets relatively near each other in the inner solar system. Yet carbon dioxide composes over 90% of the atmospheres of both Mars and Venus, whereas this gas accounts for only 0.03% of our atmosphere. Why are even these neighboring planets so different? Could their atmospheres come from different sources? To find out we must first look at our own atmosphere in more detail.

Earth is the only planet in the solar system with abundant molecular oxygen in its atmosphere. This life-giving gas composes 21% of our air, compared with nitrogen's 78%. When we emphasize oxygen's vital role for life on Earth we are being rather chauvinistic, however. It is animal life that depends on O_2. For the green plants that produce it, this gas is simply a waste product. No green plants, no oxygen; no oxygen, no animals. It is a very simple equation. If the green plants were wiped out by some giant catastrophe today, the oxygen in our atmosphere would disappear in just a few million years, as would all animal life.

So that is why only Earth has oxygen; ours is the only planet with abundant photosynthetic plants. What about the other gases in our atmosphere? For example, where does the nitrogen come from? You might suspect the Earth's volcanoes. After all, they typically produce a lot of gas when they erupt. This has been happening ever since the Earth formed, 4.6 billion years ago, so perhaps the ultimate source of the Earth's nitrogen is the rocks that compose most of the planet. Melt those rocks to make the lava in volcanoes and you will break down the nitrogen compounds they contain, producing nitrogen gas that will then enter the atmosphere. This is an appealing idea and was considered at one time to be a good explanation for the origin of atmospheric gases – not only nitrogen but water, carbon dioxide, and the noble gases neon, argon, krypton, and xenon.

A major difficulty with this idea emerges as soon as one tries to identify the minerals from which the nitrogen would come. It turns out that there is no rich source of nitrogen in rocks. The nitrates that do exist have been extracted from the atmosphere through the action of organisms or lightning discharges. Nitrates in any form are sufficiently rare that production of artificial fertilizers containing nitrogen is an important industry on Earth.

We find a similar story when we investigate the origin of Earth's carbon dioxide. Almost no carbon minerals occur in nature, diamonds and graphite being notable exceptions. Instead, we find deposits of oil, natural gas, coal, and huge amounts of carbonate rocks (Fig. 3.2) all of which were created by living organisms.

Figure 3.2 The White Cliffs of Dover, England, consist of a huge deposit of calcium carbonate ($CaCO_3$). The CO_2 that is now a component of these rocks was once part of our atmosphere. If we recovered all the CO_2 from all the carbonate rocks and oxidized all the carbon in oil and coal deposits, we would have a CO_2-dominated atmosphere 70 times more massive than the one we enjoy today.
Credit: UK-Photo.

We can now see the reason for the great differences between the composition of our atmosphere and those of Mars and Venus. In the absence of life (and liquid water, which can also form carbonates), the Earth would have a carbon dioxide atmosphere almost as thick as that of Venus. Most of the missing CO_2 on Earth is bound up in carbonate rocks. There is no liquid water or life on Venus, hence its thick, CO_2-dominated atmosphere. On Mars, we also find no liquid water on the surface today, and strong evidence that this condition has persisted for the last 3.5 billion years. This explains the persistence of CO_2 in the Martian atmosphere, but why is that atmosphere so thin? We will return to this question below.

Frigid worlds, atmospheric evolution, and cosmic time travel

We can learn another lesson about atmospheres by studying Titan, Triton, and Pluto. These three small bodies in the outer solar system all have atmospheres dominated by nitrogen, with methane and carbon monoxide (CO) as minor constituents. We certainly do not expect life on these distant, frigid worlds, so we are not surprised to find no oxygen; but why methane and not CO_2? We can find the answer by looking at Titan, the closest of the three. The reason is Titan's low temperature that acts in two ways. First, CO_2 would freeze out on the surface. Second, the surface is so cold (94 K = −290 F) that water vapor cannot exist in the atmosphere, even though ice, not rock, is the dominant material in Titan's crust. Without water vapor and green plants, there is no readily available source of oxygen that would convert methane to carbon dioxide, so this primitive hydrogen-rich gas is preserved. On Earth, Mars, and Venus, any methane that was originally present has been converted to CO_2 long ago. Life on Earth once again produces an anomaly. A certain class of bacteria living primarily in swamps and the guts of grass-eating animals continues to generate methane, but this gas is continuously oxidized to CO_2.

We can now see that even if a planet has enough gravity to maintain an atmosphere, the composition of that atmosphere can change with time. Ultraviolet light from the Sun has enough energy to break molecules apart. The fragments can then combine with other fragments to produce new molecules with a different chemical composition. This can not happen with giant planets, where the great excess of hydrogen will always maintain a steady state of hydrogen-rich gases. But on smaller planets hydrogen can escape into space, so, when methane is broken apart, the hydrogen escapes and the carbon can combine with oxygen to make carbon dioxide, if there is a source of oxygen available.

Titan, Pluto, and Triton are fascinating exceptions to this because they are so cold. Even though hydrogen is continually escaping into space, the small amount of CO_2 that is produced by the oxygen delivered by ice crystals colliding with these worlds simply freezes out on their surfaces. Hence these small worlds are literally "frozen in time." Exploring them allows us to travel back through time to study primitive environments that may contain clues to the chemical evolution that preceded the origin of life on the early Earth (Fig. 3.3).

Nearer to the Sun, things were very different. On any rocky planet close to its star in any planetary system, water vapor will be present in the atmosphere and hydrogen-rich gases will ineluctably be converted to oxidized compounds as hydrogen escapes, whether or not life develops. This is why CO_2 is so prevalent on Mars, Venus, and Earth (where most of it is now buried as carbonates). The only exception to this rule would be methane and other hydrogen-rich gases that are produced by life itself. Indeed, Jim Kasting of the University of Pennsylvania has suggested that biogenic methane may have helped to warm the early Earth through an atmospheric greenhouse effect. But how did the carbon, water, and nitrogen reach these planets in the first place?

Figure 3.3 An artist's conception of the Huygens Probe descending through Titan's atmosphere in January 2005. Saturn appears dimly in the background with the Cassini orbiter to the right. The probe's conical heat shield has been jettisoned and is hurtling towards us. Behind it, the probe descends on its parachute. The surface of Titan is decorated with pools of liquid hydrocarbons amid cliffs of water ice. Exploring this mysterious satellite will be akin to traveling back in time, as primitive, hydrogen-rich conditions are preserved in its frigid environment.
Credit: ESA.

The sources of atmospheres: problems with meteorites

If the rocks were not the original sources of the nitrogen and carbon that we now find as N_2 and CO_2 in our atmosphere, what was the origin of these elements? A popular early idea was to assume that the Earth began its existence with an envelope of gases that it captured from the solar nebula, the original assemblage of gas and dust from which the Sun and all the planets formed. The light gases would then have escaped, leaving us with the atmosphere we find today after we account for the effects of life and chemical weathering that we have just described. The difficulty with this idea is that the relative abundances of the gas-forming elements on Earth are very different from those in the solar nebula, which were identical to the abundances we now find in the Sun. There is too little nitrogen on Earth, by about a factor of five, and thousands of times too little neon, argon, krypton, and xenon, the noble (chemically inert) gases.

Evidently we need some other external source for atmospheric gases. The tradition has been to assume that meteorites that bombarded the

Figure 3.4 This 1.2 kilometer diameter crater on the plains of northern Arizona was formed by the impact of a 300,000 ton meteorite some 50,000 years ago. Impacts on both the Earth and the Moon by far bigger meteorites during the first 700 million years of Earth's history were very common, leaving the plethora of craters we still see on the Moon's surface. On the Earth the craters have been erased by erosion and plate tectonics, but some of the volatiles these meteorites delivered may still be in our atmosphere.
Credit: NASA.

early Earth in great numbers brought in most of the volatile elements and compounds, including carbon, nitrogen, and water (Fig 3.4). There has always been a problem with this idea, however. In the meteorites, the abundances of the noble gases krypton and xenon are approximately equal, whereas in the Earth's atmosphere, xenon is only one-twentieth as abundant as krypton. For many years, scientists assumed that this "missing" xenon must be trapped on or below the Earth's surface in rocks such as shales, or possibly in ice. However, attempts to find this hidden reservoir have failed; the xenon simply is not there.

Furthermore, the xenon in the meteorites appears to be fundamentally different from the xenon in our atmosphere. Xenon is a very heavy element: the nucleus of a xenon atom contains over 120 protons and neutrons (ordinary hydrogen has a nucleus consisting of a single proton). In such a massive nucleus, it is possible to vary the number of neutrons over a considerable range and still have a stable configuration that anchors a family of atoms with virtually identical chemical properties. The members of this family are called isotopes. Hydrogen has two stable isotopes, xenon has nine. This is how we can distinguish meteoritic xenon from the xenon in the Earth's atmosphere: the relative abundances of these different isotopes in the xenon found in meteorites are distinctly different from the abundances in our atmospheric xenon. These are two very different gases. Thus the Earth's inventory of volatiles differs in this essential respect from the inventory that meteorites could supply.

Why is this so important? After all, xenon is only a tiny trace constituent, less than one part per million of the gases we find in our atmosphere today. The reason we care about this tiny trace is that xenon, like all of the noble gases, is chemically inert. It is also very heavy, with three times the molecular weight of CO_2. Thus it will simply accumulate in a planetary atmosphere, neither escaping into space nor combining with other elements to form compounds (like the carbonates) on a planet's surface. The xenon we find in Earth's atmosphere today should be virtually identical to the xenon that was originally delivered to the planet, unless there was an extraordinary early episode of atmospheric escape driven by a huge outward flow of escaping hydrogen. In either case, any model for the origin and evolution of a planetary atmosphere must satisfy the constraints imposed by observed abundances and isotope ratios of xenon and the other noble gases.

The sources of atmospheres: icy planetesimals?

If meteorites cannot supply the right mixture of these critical gases, how about the comets? (Fig. 3.5). Comets, or icy planetesimals as they are more generally called (we shall use the two terms interchangeably), are the low temperature equivalent of the rocky asteroids and meteorites – icy debris remaining from the formation of the planets and satellites. This debris has left a record in the form of countless impact craters on solid surfaces throughout the solar system. Almost all of these craters were formed as the system "cleaned itself out" during the first 700 million years of its history, with a few random impacts occurring even today. Thus we are sure that icy planetesimals struck the Earth, Mars, and Venus during the early history of the solar system. How much mass was delivered by these impacts and what frozen gases and other compounds were carried in with the ice?

What we know about the composition of comets is certainly consistent with the idea that they, rather than the meteorites, could have been the major source of the atmospheres of inner planets. Like the Earth's complete inventory of gases, comets appear to be deficient in nitrogen when compared with solar abundances. Thus comets would supply the carbon and nitrogen we find on Earth today in the right

Figure 3.5 In this picture of Comet Hale-Bopp, we see the bright, white coma of dust and gas and two tails: a bright, curved dust tail and a tenuous plasma tail. The plasma tail is composed of ionized carbon monoxide, ions of nitrogen, water vapor, carbon dioxide, and other molecules are also present. All of these gases were stored in the icy nucleus of the comet and only began to evaporate into space when the comet approached the Sun. If comets struck the early Earth, they would have contributed these frozen volatiles to the Earth's inventory of gases.
Credit: Craig Gullixon.

proportions with the right isotopes. But the class of meteorites known as carbonaceous chondrites would satisfy these same criteria: they also carry carbon and nitrogen in the ratio found in our atmosphere. It is the meteoritic xenon that does not fit.

Unfortunately, we know nothing about xenon or the other noble gases in comets, despite the great success of the missions to Halley's Comet in 1986. That is not surprising because we expect that the abundances of these important gases in a cometary nucleus would be so low that the instruments on those spacecraft could not have detected them. Without direct observations, we are forced to rely on laboratory simulations to demonstrate what gases an icy comet nucleus is likely to contain, assuming that it forms at some specified temperature – hence distance from the Sun – in the outer solar nebula.

A series of such experiments has been carried out by Akiva Bar-Nun and his colleagues at the University of Tel-Aviv in Israel. The basic procedure involved the deposition of a pre-determined mixture of gases and water vapor on a "cold finger" inside a vacuum chamber. The temperature of the cold finger represents the temperature at which the ice grains that compose a comet nucleus would form. The gas mixture was chosen to duplicate the abundances of gases expected to be present in the interstellar cloud from which the solar system originated, thus these are also the abundances in the original solar nebula. After the ice was formed, the chamber was evacuated again and the cold finger was gradually warmed up, causing gases in the ice to escape. The proportions of the gases that had been trapped in the ice were then measured.

The results from this work support the idea that comets must have been an important source of the volatiles we now find on the inner planets. At the slow rates of water vapor deposition that would characterize formation of ice grains in an interstellar cloud or in the outer reaches of the solar nebula, argon, krypton, and xenon were trapped in ice exactly in their original proportions at ~10 K. At a temperature of 25 K, argon is trapped much more poorly than krypton and xenon, such that the proportions of these three gases in the ice are entirely consistent with the abundances we find in the atmospheres of Mars and Earth. In other words, if the krypton and xenon in the Earth's atmosphere were brought to our planet by ice that formed at a temperature near 25 K, we would find the abundances of these two gases to be very similar to the ratio of 20 to 1 that we find in our atmosphere today, rather than the nearly equal abundances carried by the meteorites.

The temperature of 25 K may be significant because this is the same temperature deduced from studies of the spins of hydrogen nuclei in cometary water and ammonia, and is a temperature measured in interstellar clouds. It suggests the preservation of interstellar ice grains in comets.

Solar Composition Icy Planetesimals (SCIPs): a new type of icy planetesimal

There is some new evidence indicating that interstellar abundances were in fact preserved in icy planetesimals in the early solar system. This evidence comes from an investigation of the composition of the atmosphere of Jupiter carried out by the mass spectrometer on the Galileo Entry Probe (Fig. 3.6). The favored model for the origin of Jupiter starts with a solid core built by the accumulation of icy planetesimals.

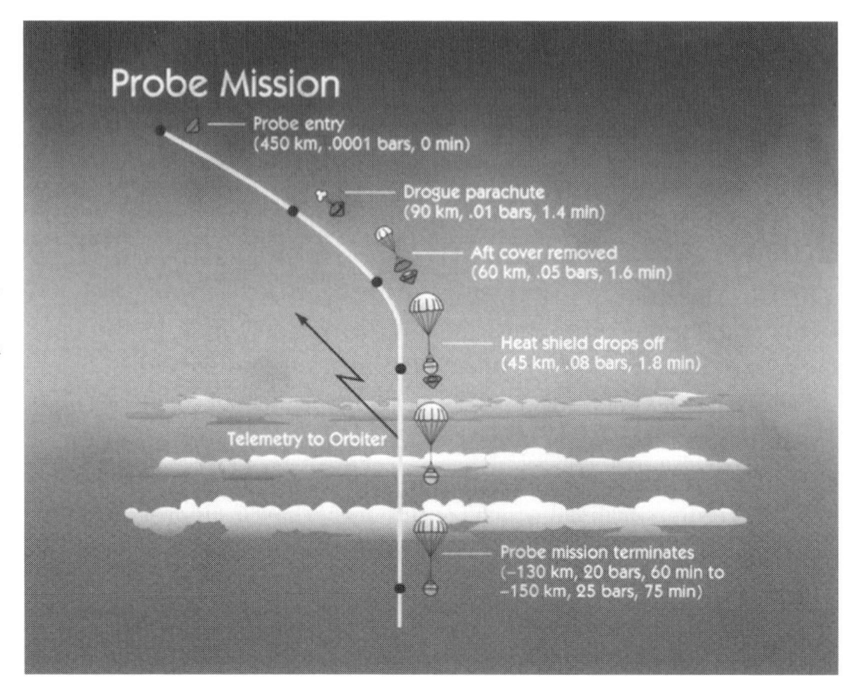

Figure 3.6 This NASA artwork shows the trajectory of the Galileo Probe in the atmosphere of Jupiter on December 7, 1995. As it descended through the predominantly hydrogen–helium atmosphere, it measured elemental abundances and isotope ratios of noble gases and other species that are helping us to understand the contribution of icy planetesimals to the origin of this giant planet and to the origin of our own planet's atmosphere.
Credit: NASA.

When this core grows to a size of 10 to 15 times the mass of the Earth, it can attract hydrogen, helium, and all the other gases from the surrounding solar nebula gravitationally. The resulting increase in mass enables the attraction of more material, including additional planetesimals, and soon a giant planet is formed. The atmosphere of this giant should then consist of a mixture of gases from two distinct reservoirs: the gases contributed by icy planetesimals – those that built the core and those that subsequently impacted the growing planet – and the gases attracted directly from the solar nebula. This should lead to an enrichment of the gas-forming elements that were brought in by the planetesimals, compared with solar nebula (hence solar) abundances.

The icy planetesimals participating in this process were considered to be identical to comets, which led to the expectation that nitrogen and noble gases would be deficient on Jupiter, as they are in the comets. However, the Galileo Probe found that the noble gases and nitrogen were enriched on Jupiter by the same factor of 3 ± 1 as carbon and sulfur. In other words, the icy planetesimals that contributed to the formation of Jupiter were apparently composed of a mixture of elements in solar relative abundances, deficient only in hydrogen, helium, and possibly neon. These icy building blocks of Jupiter can therefore be called Solar Composition Icy Planetesimals, or SCIPs. If they also formed the cores of the other giant planets, the SCIPs must have been the most abundant form of solid matter in the early solar system.

SCIPs are clearly very different in composition from the comets we know, suggesting that these planetesimals must have been formed at temperatures below 25 K so they could capture solar amounts of nitrogen and argon. This requirement in turn suggests the incorporation of interstellar icy grains in comets, as no one has yet been able to develop a scenario that would allow the requisite mass of low temperature icy planetesimals from the Kuiper Belt in the outer reaches of the solar system to migrate into 5 AU to build Jupiter.

A second result from the Galileo Probe provided a valuable insight into the origin of the nitrogen we breathe on Earth. Jupiter's nitrogen exhibits a distinctly different ratio of the two abundant isotopes from the ratio found on Earth. The difference is exactly what one would expect if Jupiter's nitrogen had been delivered to the planet as N_2, while Earth's nitrogen arrived in the form of nitrogen compounds such as ammonia or hydrogen cyanide.

The presence of this uniform enrichment of heavy elements on Jupiter, together with the evidence that the nitrogen on Jupiter arrived there in the form of N_2, puts an interesting constraint on models for the giant planet's formation. Evidently it would be very difficult for the planet to originate through a gravitational instability in the original solar nebula, causing the direct condensation of nebular material to form the planet. In that case, Jupiter should now have the same composition as the Sun (which certainly formed directly from the nebula), whereas it obviously does not. Hence the alternative scenario for making Jupiter – building up a massive, icy core that then causes the gravitational collapse of surrounding nebular gas – seems much more likely.

Now we can begin to put some important pieces of the puzzle together. The major reservoir of nitrogen in the solar nebula was N_2, which is difficult to trap in ice because of its high volatility. That is evidently the reason why nitrogen (the element) is deficient in comets. But ice forming at very low temperature can capture N_2, which is why SCIPs, unlike ordinary comets, exhibit the same enrichment of nitrogen as carbon or sulfur. Ordinary comets hitting the Earth would bring a small amount of nitrogen in the form of compounds. These nitrogen compounds were rapidly broken apart by ultraviolet light from the Sun, and the liberated nitrogen atoms combined to make the N_2 in our atmosphere. That is what you are breathing right now! The same thing should have happened on Mars, Venus, and Titan.

Why didn't SCIPs hit the inner planets? Perhaps the core of Jupiter formed well before the inner planets came together, as many scientists now think, and the ordinary comets formed later. There is a hint of a contribution of SCIPs to the noble gas abundances on Venus, but we need a new mission to that planet to measure those gases more carefully before we can be sure.

The sources of atmospheres: a rocky component?

The comets did not contribute all the gases we find on the inner planets, however. There is evidence in both the Martian and terrestrial atmospheres for another source – probably the rocks that constitute the bulk of these planets. This evidence consists in part of the exact proportions of noble gases in the planetary atmospheres. The abundances of the gases trapped in laboratory ice do not duplicate the values in the Earth's atmosphere. Instead they establish one end of a "mixing line" that connects a hypothetical reservoir within the planet with the values found in the ice. The abundances of noble gases in the atmospheres of Mars and Earth fall along this line. The mixing line can then be interpreted as representing the mixture of these two components – one coming from the interior of each planet and the other coming from icy planetesimals that impacted the planets long ago. The resulting mixture is what we observe in the planetary atmospheres today. The noble gases found in meteorites from Mars fall along this same mixing line indicating that they represent a varying mixture of gases from the Martian atmosphere and gases from the planet's interior.

Another reason for invoking a second source of volatiles comes from an examination of the inventory of carbon and oxygen in the Earth's crust and atmosphere. The ratio of abundances of oxygen to carbon on Earth is roughly two to three times greater than the value that would be supplied by either comets or meteorites. An additional source of oxygen, most likely water (H_2O), is needed.

The smoking gun that forces us to require that atmospheric gases must come from more than one source is the isotopic composition of water on Mars and Earth. Recall that hydrogen has two stable isotopes, the common form with a single proton in the nucleus, and the heavy isotope known as deuterium, in which that proton is joined by a neutron, giving this isotope twice the mass of its more abundant brother. Despite this difference in mass, however, the behavior of these two isotopes in chemical reactions is virtually identical, as the chemistry is determined primarily by the single electron that both forms of hydrogen possess. Thus the proportion of deuterium to ordinary hydrogen (abbreviated as D/H) in any hydrogen compound can tell us something about the history of the hydrogen that we cannot deduce from the compound itself. We can therefore use this property to investigate the source(s) of water on Earth and Mars.

We discover immediately that the Earth's oceans cannot consist simply of melted comets. Thanks to the recent appearance of two bright comets – Hyakutake and Hale-Bopp (Fig. 3.5) – allowing microwave observations from Earth – and the highly successful Giotto mission to Halley's comet in 1986, we have three determinations of HDO/H_2O in cometary ice. These studies all gave the same result: D/H in comets is about twice the value found in our oceans. So we need another source of water – with lower D/H – to combine with the water from melted comets and produce the oceans.

An obvious candidate is the huge mass of the Earth's rocky interior. We know that the rocks that formed the Earth trapped volatiles before they came together to compose the planet because we can find neon captured from the ancient solar wind inside rocks that formed from magma originating deep inside the Earth. The magma reaches the surface at mid-ocean ridges and hot spots, forming rocks that we can examine in our laboratories. Unfortunately we cannot hope to find pristine water in such rocks, preserved from the time the Earth formed, owing to contamination by surface water that gets recycled through the Earth's interior by tectonic processes. (Because of the small size of its atoms and its chemical inertness, neon is not recycled in this way.) So again we have to rely on laboratory studies to help us deduce what was happening to water vapor in the solar nebula before the planets appeared.

Investigations by Dominique Lecluse and Francois Robert of the Museum of Natural History in Paris have demonstrated that water vapor in the inner solar nebula where the Earth was formed would achieve a value of D/H that could be less than half the value found in the oceans. This occurs as a result of the exchange of deuterium between H_2O and the huge amount of H_2 in the nebula, because D/H in H_2 is one-tenth the value of H_2O. Some of this isotopically modified water vapor could have been adsorbed on the silicate grains that accumulated to form the rocks that made the Earth. Even after the huge collision that produced the Moon, apparently enough water was left inside the Earth (we still see that solar wind neon!) to mix with cometary water and make the oceans. Subsequent bombardment of the Earth by comets would add water with higher D/H, producing the blend that we now find in the Earth's oceans.

Something similar should have happened on Mars as well. However, because this small planet is much less geologically active than the Earth, there has been little or no mixing of crustal reservoirs with the interior. We can thus hope to find evidence of this rocky water reservoir preserved in the interior of that planet. We already seem to be seeing the cometary contribution near the surface of Mars: the lowest values of D/H found in water-containing minerals in Martian meteorites are about equal to the value found in cometary ice, twice the value in the Earth's oceans.

The importance of impact erosion

Impacts not only bring material in, they can also remove it. The presence on Earth of meteorites from Mars is a striking example. Large impacts can even remove fractions of a planetary atmosphere, a process called "impact erosion." This process was especially effective in the case of Mars because of the planet's small mass, hence its weak gravitational field. Impact erosion during the first 700 million years of the planet's history appears to be responsible for making the present atmosphere of Mars so terribly thin. Calculations by Jay Melosh and Ann Vickery at the University of Arizona show that at least one hundred times the present mass of the Martian atmosphere must have been removed by the same early bombardment that brought in volatiles and left a record of impact craters.

Poor Mars! Its fate was sealed by its small mass (Fig. 3.7). Yet even Mars may have enjoyed episodes of high surface pressure during that early epoch of its existence. The constantly shifting balance between delivery and removal of volatiles as the ferocious early bombardment ran its course would have included intervals of time in which the atmosphere

Figure 3.7 The arid surface of Mars seen in this global view serves as a reminder that the early bombardment by icy and rocky planetesimals could bring both life and death. Once water raged across the surface of this planet, and pooled in its impact craters, but repeated impacts depleted the atmosphere so severely that no open bodies of water exist on Mars today. *Credit:* NASA, USGS.

was much denser than it is today. These episodes would have allowed the formation of the famous Martian channels and other features caused by aqueous erosion (see Chapter 4).

Tests of the model

Our answers to the question "Why does the Earth have an atmosphere and oceans?" are thus "because it is the right distance from the Sun, the right mass, and because it got some critical help from comets!" Bar-Nun and I like to call this the "icy-impact" model for the origin and early evolution of the inner planet atmospheres. It seems to be generally self-consistent. However, a model is only as good as the experiments that can test it. There is much more that we need to know before we can fully accept this view of the origin of our atmosphere.

Figure 3.8 The Rosetta Mission will arrive at comet 67P/Churyumov/Gerasimenko in August 2014. It will rendezvous with the comet and deploy a probe that will land on the surface of the comet's icy nucleus, seen at lower right. *Credit:* ESA.

The most essential test is a direct determination of abundances and isotope ratios of noble gases in comets, so we do not have to rely on laboratory simulations. We have not yet detected any noble gases in comets! What ratio of krypton to xenon do these icy messengers from the outer solar system contain? Is it similar to the value on Earth and Mars or does it resemble the meteoritic ratio? We have not yet been able to demonstrate in the laboratory that the isotopes in cometary xenon should match the pattern found in the Earth's atmosphere even though the krypton/xenon abundance ratio fits. Perhaps some vigorous, non-thermal escape process is required to convert the solar isotopic pattern of xenon to the atmospheric pattern after the gases are delivered to the planets, as suggested by Robert Pepin of the University of Minnesota. What is the pattern of xenon isotopes in comets?

The first answers will come from the ESA Rosetta mission in 2014–15 (Fig. 3.8). Rosetta will rendezvous with Comet 67P/Churyumov/

Figure 3.9 Three close-up views of cliff faces on Mars. The left and middle images show gullies cut by running water that have created aprons of debris overlapping relatively recent dunes and polygonal ground, respectively. The right-hand image shows gullies cut into dust-free ground. Over a thousand of these small gullies have been imaged, and many appear to be geologically recent. Some may even be forming now. *Credit:* NASA.

Figure 3.10 Artist's conception of the first Sample Return mission leaving Mars. The return vehicle is streaking upward into the Martian atmosphere after being loaded with samples collected by the foreground rover. *Credit:* NASA.

Gerasimenko, accompany the comet in orbit, and will deploy a probe that lands on the icy nucleus. Comet Churyumov/Gerasimenko is thought to have originated in the Kuiper Belt, whereas Halley, Hyakutake, and Hale-Bopp are all Oort Cloud comets. Thus the new data set that will come from this mission will also allow us to determine whether or not there are any fundamental compositional differences between the two cometary reservoirs. There is an important caveat that must be mentioned here, however. Comet Churyumov/Gerasimenko has spent thousands of years moving through the inner solar system between the orbits of Jupiter and Earth. Thus it has had ample opportunity to be "baked out" and may have lost all or nearly all of its most volatile constituents such as N_2. So even though it was a comet born in the Kuiper Belt, Churyumov/Gerasimenko may no longer exhibit the chemical composition expected from that low temperature origin. Nevertheless, if there really is a systematic difference between the deuterium to hydrogen ratio in Oort Cloud and Kuiper Belt comets that would allow the latter to be a source of the Earth's oceans, that difference should be preserved. Furthermore, the comet may have retained some xenon and studying those xenon isotopes will be a crucial experiment for determining the role of comets in bringing volatiles to the early Earth. Thus Rosetta promises a major advance in our efforts to understand how the Earth got its atmosphere.

And what about the oceans? How can we test the idea that the oceans are a mixture of cometary and planetary water? Here our best hope for recovering the original scenarios may lie with Mars. Rocks ejected from deep craters on Mars may contain some water from the planet's interior. That water should exhibit the low values of D/H expected in the inner solar nebula, if indeed the water locked in the interior of Mars has remained isolated from the water at the surface. The intriguing narrow gullies cut into some cliffs by subsurface water emerging from the cliff faces may also hold a clue (Fig. 3.9). That water must come from ice that has formed at some depth below the Martian surface. Is the ultimate source of this ice water that percolated up from the interior of Mars, or did the water seep down from the cometary contribution to the crust? Or is it a mixture? Retrieving some of that ice and analyzing its isotopes will give us the answer. This is another exciting goal for the Mars Sample Return missions to investigate when this ambitious program begins to bring us carefully selected rocks from the red planet sometime after 2010 (Fig. 3.10).

MICHAEL T. MELLON
University of Colorado at Boulder

The frozen landscape of Mars

Mars: yesterday and today

In 1894, intrigued by Schiaparelli's reports of "canali" on the surface of Mars, Percival Lowell and his colleagues constructed a new observatory in the mountains near Flagstaff, Arizona. In little over a month after breaking ground on the site that April, they began conducting regular observations of the red planet. Their interests were clearly focused on "canals" and their telescope included instruments for accurately measuring the positions and widths of these apparent features. Additionally, they made observations of other surface features, the changing Martian seasons, and the planet's moons. These observations contributed to improving the scientific understanding of Mars.

Lowell believed Mars to be a warm planet, where water flowed through canals supporting vegetation and even civilization. Although the idea of intelligent Martians constructing canals was limited to only a few observers, the prevailing view of that time was that Mars was very much like the Earth.

Some observers thought dark regions to be vast seas connected by waterways of natural, possibly geologic, origin. The absence of a reflected glint from the Sun and a blue-green appearance led others to believe that these areas were covered by vegetation. Dark bands surrounding the receding polar caps were perceived as wet soil, darkened by water from the spring snowmelt, possibly combined with peripheral vegetation. In 1904, John Henry Poynting estimated the surface temperature of Mars to be about 235 K (−38 C), far below freezing. However, his colleagues dismissed his result because it did not agree with the vision of a warm Mars that observers thought they were seeing in the eyepieces of their telescopes.

Today, after several decades of exploration by spacecraft such as Mariner, Viking, Pathfinder, Mars Global Surveyor, and Mars Odyssey, along with Earth-based observations using techniques not available in Lowell's day, a very different picture has emerged. Mars is a frozen desert.

Being farther from the Sun than the Earth, Mars receives less than half the amount of warming sunlight received by the Earth and is therefore considerably colder, just as Poynting had predicted. The global average surface temperature is around 205 K (−68 C), similar to wintertime in the Antarctic interior. Martian polar temperatures are much colder than any region of Earth, dipping as low as 145 K (−128 C). At the equator, under the noonday Sun, while surface temperatures may briefly reach (or even exceed) the freezing point of water (273 K), they typically average a chilly 220 K (−53 C).

In addition to these cold temperatures, the thin Martian atmosphere is extremely dry. There is barely enough water vapor to coat the ground with a paper thin layer of frost. By comparison, the atmosphere of Earth typically contains enough evaporated water to coat the entire planet's surface with a layer of rain water nearly 10 centimeters deep. Despite this dryness on Mars, the relative humidity varies from desert-like conditions to near saturation. The relative humidity is a measure of the amount of water vapor the air holds relative to how much water the air can accommodate without condensing to form clouds or frost. Near the equator, where temperatures are the highest, the relative humidity is the lowest. This same quantity of water vapor close to the pole would nearly saturate at the colder temperatures, yet conditions would still be considered dry because very little water vapor is actually present.

If an astronaut were to set a glass of water on the Martian surface near the equator, it would rapidly begin to evaporate, and possibly boil. (The boiling point on Mars is very near the freezing point because of Mars' low atmospheric pressure, about 150–200 times lower than Earth's sea-level air pressure.) In a few moments the water in the glass would form an icy crust, primarily through evaporative cooling, in much the same way sweat cools the skin when it evaporates. The ice surface would lose water more slowly than the liquid water surface, by the process of sublimation (changing directly from solid to gas), giving the rest of the water time to freeze. Eventually, even at this slower rate of loss, all of the water ice would turn to vapor and be carried away in the Martian wind, leaving an empty glass. If the astronaut placed a similar glass of water at a higher latitude, near the Martian pole, it also would first evaporate and freeze, much like the glass near the equator. Then, as the water turned to ice and reached thermal equilibrium with its surroundings, it would stop sublimating and would remain as a glass full of ice.

Under these cold and dry Martian conditions any water present at or near the surface of Mars must be in the form of ice. And ice, occupying a variety of niches, has a fundamental impact on the Martian environment and will influence the humans who may some day go there. Interaction between ice and the atmosphere is a basic component of the Martian climate. The possible existence and evolution of Martian life in a form similar to terrestrial life may be strongly influenced by the availability of ice as a source of moisture. Ice will be a readily accessible and most abundant reservoir of water for future human settlement. Indeed, the availability and geographic distribution of ice may put constraints on the design and location of future human settlements on Mars. Much of the diverse character of the Martian landscape owes its existence to the presence of ice and of melt-water from that ice. While some aspects of ice in the Martian environment are still debated among scientists, ice has played and will continue to play a central role in shaping the landscape, the climate and the actions of future astronauts.

Polar deposits

Since the earliest telescopic observations of Mars, the polar caps have been the most striking features (Fig. 4.1). Seen at both north and south poles, these bright white patches have long been observed to grow and shrink with the changing Martian seasons. In Lowell's day these features were generally considered to consist of water ice that melted in the Martian spring. Although some observers suggested carbon dioxide ice as an alternative explanation, this idea was dismissed as it would require temperatures too cold for liquid water, which was thought to exist over much of the planet's surface.

In reality, the polar deposits on Mars are far more complex than previously believed. Seasonally, a layer of carbon dioxide ice – the gas that makes up most of the thin Martian atmosphere – blankets the surface during the long polar night, sublimating in spring and summer. Beneath this transient layer persists a thick and extensive deposit of ice – the bright white cap is merely the tip of the iceberg. Surrounding both ice caps are the polar layered deposits, dirty looking deposits believed to be composed primarily of ice intermixed with some dust and sand.

Figure 4.1 Mars seen from Earth. This image taken from Earth orbit shows contrasting light and dark surface features, but the most striking feature is the bright north polar cap. A thin layer of seasonal carbon dioxide frost left over from the previous winter blankets the polar surface.
Credit: NASA, Hubble Space Telescope, image courtesy of Dr Steve Lee.

The vast expanse of the northern polar ice cap is nearly 1,000 kilometers across (and 20% larger when including the surrounding layered deposits), and totals about 3–4 kilometers deep at its center (Fig. 4.2). The southern polar ice cap is about one quarter the size of its northern counterpart. However, the southern polar ice deposit, including the layered deposits, comprises a considerably larger area, though similar in thickness relative to the northern deposits. The highly irregular cratered terrain around and beneath the southern deposit makes it difficult to estimate its total volume. In fact, the south cap has the added peculiarity of not being centered on the pole. This polar cap offset is thought to be due to the topographic influences of the South Polar Basin, a large and ancient impact-generated basin and ring of mountains near the south pole. Scientists are unsure about the volume of water present within the polar ice. Assuming the entire mass of the

Figure 4.2 Residual ice of the north polar cap. In the summer, sublimation of all of the seasonal carbon dioxide frost exposes the thick residual water ice cap. Valleys clearly visible in these images, several kilometers wide, spiral outward from the cap's center.
Credit: NASA, Viking Orbiter.

deposit is comprised of water, both poles could contain enough water to create a layer of water about 20–30 meters deep if spread over the entire surface of the planet.

Earth's massive ice sheets make for an interesting comparison with the ice deposits on Mars. The Antarctic ice sheet is about 4,500 kilometers across and amasses 3.5–4 kilometers of ice at its thickest point. The Greenland ice sheet measures about 900 kilometers wide by 2,100 kilometers long with about 2–3 kilometers of ice piled on top. The continental glaciation on Earth represents a similar thickness of ice deposit to Mars, but Earth's ice sheets cover a much larger area than the Martian polar deposits.

At both poles, deep valleys that cut into the deposits show alternating light and dark bands or layers (Fig. 4.3), hence the term "polar layered deposits." These layers are believed to be layers of ice and dust or layers of high and low dust content within the ice. Individual pairs of layers are primarily about 10–50 meters thick, with thinner 1–3 meter layering evident in some higher resolution images. These layers are thought to be a record of periodic climate change coincident with enormous

Figure 4.3 Polar layers. Valleys cutting into the polar cap and surrounding ice deposits show numerous light and dark bands and parallel ridges on the valley walls. The bands suggest that the deposits are made of alternating layers of ice and dust or layers of clean ice and dirty ice. Large-scale layering occurs at about 10–50 meter intervals, while smaller 1–3 meter thick layers are evident in some higher-resolution images. Image width is 2.4 kilometers. *Credit:* NASA, Mars Global Surveyor.

oscillations in the Martian orbit. As the climate oscillates, alternate periods of deposition or erosion of ice and cycles of atmospheric dust storms could ensue.

Another unusual aspect of the polar deposits is that these valleys spiral outward from the center of the cap. These spirals run counterclockwise in the north (Fig. 4.2) and clockwise in the south. The origin of the valleys themselves probably relates to preferential heating and the subsequent sublimation of ice from the equatorward (therefore sunward) facing slopes. Likewise, poleward facing slopes remaining in the cold shadows may trap some of the escaping moisture. While the reason for the spiral pattern remains a mystery, some scientists suggest that local wind patterns may play a role.

The polar caps and surrounding layered deposits are some of the youngest terrains on Mars. The ages of these icy surfaces can be estimated by counting impact craters, a technique known as crater age dating. Since impact craters are created occasionally but regularly throughout the histories of all the planets, their abundance dates a surface. A surface pitted by many craters developed a long time ago, while a surface with few craters is a newcomer to the geologic scene.

While far from exact, the absolute age can be approximated by estimating the rate at which Mars has been bombarded.

The north polar deposit is devoid of observable impact craters. Therefore, this deposit's surface is very young, perhaps younger than ten million years. The south polar layered deposits have just a few craters and may be slightly older, but the south polar cap itself contains no impact craters and is also considered very young. Although polar ice deposits may have endured throughout Martian history, these young ages provide evidence of an active polar landscape where resurfacing wipes the cratering record clean.

Seasons bring change

As on Earth, a tilt in the Martian polar axis causes seasons. Like a spinning top, a planet prefers to tilt its poles in one direction as it revolves around the Sun. This polar tilt, called the obliquity, is about 23.5 degrees for the Earth and is 25.2 degrees on Mars. This angle causes one hemisphere to tilt away from the Sun for half the Martian year (a Martian year being about 1.9 Earth years) and to tilt toward the Sun for the other half. The hemisphere tilted away from the Sun receives less solar energy, or none at all at the higher latitudes, relative to the hemisphere tilted towards the Sun. Reduced sunlight causes surface temperatures to fall and that hemisphere experiences a winter. Half a Martian year later the Sun rises higher in the sky, the surface warms, and the season changes to spring and summer. Of course, the opposite hemisphere would experience the same seasonal changes, only half a year later.

Changes in the Martian seasons bring dramatic changes to the polar deposits. During the winter, polar temperatures fall low enough to condense CO_2 (carbon dioxide) gas to a solid frost known on Earth as dry ice. CO_2 is the atmosphere's primary gas, accounting for about 95% of the Martian air. Although the atmosphere is exceedingly thin, as much as several meters of CO_2 frost can accumulate on the ground during the polar night, causing the atmospheric pressure to drop by 25% over the entire globe. Some of this seasonal ice may be deposited as a carbon dioxide snow, where dry ice encrusted dust grains gently settle to the surface. Some of this ice may also be directly deposited on the surface like a frost deposited inside of a kitchen freezer.

In either case, the seasonal coating of CO_2 frost extends equatorward from the pole to about 40 degrees latitude each year. This winter cycle compares with what happens on Earth when seasonal snowfall blankets a similar range of latitudes. However, the Martian seasonal frosts do not melt when the warmer spring temperatures return, but instead sublimate back to atmospheric gas.

In the north, during the Martian summer, all of the seasonal CO_2 frost will sublimate and expose the residual ice cap. The surface of the northern residual ice cap is known to be water ice. The temperature of this bright icy terrain as measured by orbiting spacecraft is too warm to sustain CO_2 ice, indicating that all the CO_2 frost has returned to the atmosphere. At the same time, the atmospheric humidity has been observed to increase dramatically over the north polar region. As the increased summer sunlight heats the residual water ice surface, it sublimates into the air, awaiting the Martian winds to distribute it over the globe.

Conversely, in the south, not all the seasonal CO_2 frost disappears each summer. A residual CO_2 ice surface persists through the Martian summer, to be covered by fresh carbon dioxide frost in the following fall and winter. However, this residual CO_2 ice covering left behind in the summer is believed to be fairly thin, concealing a water ice substrate beneath. Unusually high humidity has been observed some years over the southern pole in its summer – a regular occurrence during the northern summer. Such an abundance of water vapor would not have been possible had residual summer CO_2 frost still blanketed the surface. These observations also suggest that climate on Mars can vary slightly from year to year.

Shapes in the polar landscape

In the polar summer when the seasonal CO_2 frost has gone, the vast windswept terrains of the ice caps are exposed. The surfaces at each of the poles differ dramatically. Small pits and knobs, a few meters in size, dominate the northern polar cap (Fig. 4.4). Nearer to the polar valleys, this knobby terrain gradually makes a transition down to the exposed layers. Sublimation-driven ablation of water ice exposed to the summer Sun and erosive power of the Martian wind may be responsible for this hummocky texture that extends for hundreds of kilometers.

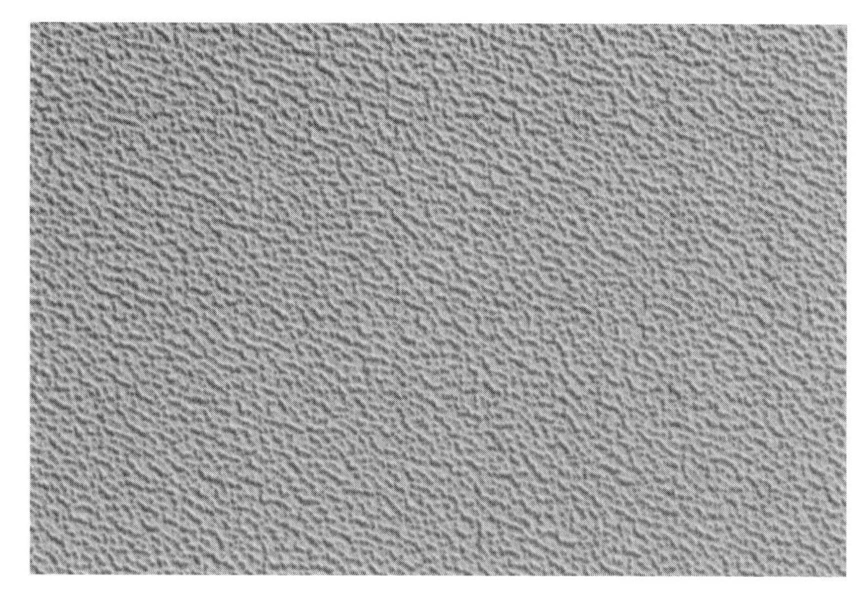

Figure 4.4 The northern polar cap surface. This windswept terrain is peppered by several meter-sized knobs and ridges thought to be formed by the ablation of water ice. Sunlight heating the polar surface drives sublimation and erosion of the ice. Wind may pile ice grains and snow adding to the bumpy texture. Scene width is about 1.8 kilometers.
Credit: NASA, Mars Global Surveyor.

The southern polar cap is much more unusual. Marked by circular depressions, narrow parallel troughs, and numerous fractures (Fig. 4.5 and Fig. 4.6), this surface has apparently evolved in a distinctly different way from the north. Nearly circular pits, a few hundred meters in size with sharp boundaries, are common, indicating that several meters of material (polar ice) has been removed in a very local area. Less-defined shallow depressions may hint at the early stages of their formation. Possibly, buried accumulations of CO_2 ice become too warm and sublimate through an overlying lag of water ice and dust; the lag then collapses into the spot once occupied by the CO_2 ice. In the polar summer, the Sun remains low in the sky circling throughout the day near the horizon. A subtle depression might be warmed more effectively than its surroundings, enhancing the depression until a deep pit forms.

More difficult to explain are elongated parallel troughs, resembling a human fingerprint (Fig. 4.6). Individual troughs are tens of meters wide and spaced roughly a hundred meters apart. Sublimation of ice, perhaps on the sunlit side of each trough is most likely responsible, but the regularity of the patterns remains a mystery. Although similar processes of cooling in the polar night and solar heating in the day drive cycles of ice condensation and sublimation at both poles, the resulting landscape at north and south poles is remarkably different.

Figure 4.5 Circular and arcing pits in the southern ice cap. Several meters of icy material, possibly CO_2 ice, has eroded away within these very localized circular, hundred-meter-sized depressions. In some areas, thin layering can be seen in the walls. Polygonal patterns 20–50 meters in scale are common on the uneroded surfaces. Scene width is about 2.3 kilometers. *Credit:* NASA, Mars Global Surveyor.

Figure 4.6 Thumbprints on Mars. These parallel ridges resembling a fingerprint are common on the ice at the Martian south pole. Individual ridges and troughs are spaced about 100 meters apart. Layers are frequently exposed within the troughs. Sublimation along the trough walls could be responsible for this unusual landscape. Image width is about 4.5 kilometers. *Credit:* NASA, Mars Global Surveyor.

Deep in the ice cap

While the surface of the polar ice deposits, at least in the north, is clearly made of water ice, scientists have debated whether the deep interior of the deposits may be composed of some other form of ice. What might an astronaut find drilling deep into the polar landscape? If CO_2 frost can condense on the surface, could it persist at depth?

Some fraction of the polar deposits might consist of buried CO_2 ice, trapped as small inclusions by water ice accumulating on the polar surface. If this CO_2 is released into the atmosphere, it could have a pronounced effect on the climate (CO_2 is the greenhouse gas that gives Venus its scorching temperatures). Likewise, sequestering CO_2 in the polar deposit and removing it from the atmosphere would lessen any greenhouse warming produced during a past age and give the Martian climate an icy chill.

An unusual type of ice that may also exist in the polar deposits is a crystalline combination of CO_2 and water called CO_2 clathrate hydrate. This special solid phase of water requires one CO_2 molecule for every six water molecules to maintain its crystalline structure. This form of ice can only exist under certain pressure and temperature conditions. Although clathrate has never been detected on Mars, favorable conditions prevail seasonally at the polar surface and continuously at a depth of a few meters under the polar water ice. Clathrate hydrate is of interest, not only because it contains water, but it can store large quantities of CO_2.

How much CO_2 could be stored in the polar deposits? Based on the thermodynamic stability of the different ice species and the topography of the polar deposits, only a small amount of CO_2 ice or CO_2 clathrate hydrate could be present. Thus, most of the immense mass of the polar deposits is a reservoir of water ice. Still, if all of this CO_2 were released into the atmosphere, the pressure could increase by a factor of 10 or more. This additional atmospheric gas would have a dramatic effect on the climate, raising the surface temperature and increasing the atmosphere's ability to transport heat and windblown dust and sand. Weather on Mars would be very different from the Martian weather we observe today.

The sky above

Water in the atmosphere of Mars is an important component of the Martian climate and the cycles of ice on its surface. Although water is mainly a vapor in the atmosphere, it serves as a conduit moving ice from one location to another. Water and CO_2 ice clouds form high above the ground or as an ice fog closer to the surface in valleys and canyons. The interaction with the polar water ice deposits provides the entire atmosphere with the little moisture it contains.

In the summer, when the blanket of seasonal CO_2 frost has dissipated, the increased sunlight warms the polar water ice surface. The warmer ice sublimates more rapidly than it would at colder temperatures and the sublimated vapor rises to humidify the air above the pole. The general circulation of winds around the planet carries this water vapor to other locations.

Eventually, a small amount of this water will be carried to the opposite (winter) pole. There, the extremely cold surface, covered with CO_2 frost, will trap water by condensation and incorporate it within the seasonal CO_2 deposit. Much of the sublimated water will linger near the summer pole it came from, and will be re-deposited back on that pole half a Martian year later when winter returns. If both poles manage to lose their seasonal CO_2 frost covers during their respective summers, water ice will be transferred back and forth between the polar deposits. However, if the south pole retains a residual CO_2 frost cover in most years, as has been observed, this cover will prevent the warm temperatures and ice exposure needed to sublimate water ice and to return water vapor to the north pole. Hence a net accumulation of ice from year to year will occur at the south pole at the expense of ice mass from the north pole. Which scenario is an accurate description of the current Martian climate remains to be discovered.

Ice clouds, fogs, and hazes can develop when the relative humidity reaches 100%. Clouds have been observed on Mars for many years. Dust storms form yellow or reddish clouds seen in the telescope, where atmospheric winds lift the rusty surface dust into the sky. Bluish and white clouds and hazes have also been observed with regularity (Fig. 4.7). These clouds form by condensation of water ice, and possibly CO_2 ice, within the atmosphere.

Many of the water ice clouds appear stationary, occurring over mountains, seasonally at the equator, or over the poles in winter. Mountain clouds, called orographic clouds, can remain stationary for many days. These clouds are comprised of water vapor lifted to higher altitudes and carried by the wind in a trek over a mountain range. As air is raised to higher altitudes it cools and a cloud forms; as the air falls again at the end of its mountain journey, the air warms and the ice crystals comprising the cloud sublimate back to a vapor. Although the individual

Figure 4.7 Clouds on Mars. Water ice clouds occur regularly during the Martian early morning hours.
Credit: NASA, Mars Pathfinder.

water molecules in the cloud differ from day to day, or even from hour to hour, the cloud appears to remain stationary. This sort of cloud also occurs commonly in mountainous regions on Earth.

Relatively stationary clouds occurring in the winter over the polar regions are collectively referred to as the polar hood. These clouds are primarily composed of water ice crystals. Conditions are favorable for some CO_2 ice clouds to be incorporated as well. Although orbiting spacecraft have detected CO_2 ice clouds on Mars, the frequency with which such clouds occur is still unknown.

Many ice clouds are dynamic in character and move across the planet's surface just as frontal systems move over the surface of the Earth and give us our changing weather. Unlike Earth, though, ice clouds on Mars remain thin and wispy, because the dry Martian atmosphere simply does not contain sufficient water to form thick clouds. Nevertheless, clouds are an important part of the Martian climate.

The permafrost below

Water can also be found as ground ice, buried ice within the Martian permafrost. The term permafrost describes the region within the soil where the temperature remains permanently below the freezing point of water, 273 K. On Earth, this region is restricted geographically to the far north and south, for example in northern Alaska and Canada, Siberia, and Antarctica, as well as in some high mountain locations at lower latitudes.

On Mars, permafrost is found globally, since subsurface temperatures are well below freezing everywhere. Only at depths of several kilometers or more will temperatures rise above freezing due to geothermal heat from deep within the interior of Mars. Because of the planet-wide occurrence of permafrost on Mars, scientists sometimes refer to this region between the surface and a depth of several kilometers as the cryosphere.

Just as on Earth, the ice content of the permafrost will depend on the specific moisture conditions for each region. On Mars, these conditions are controlled by soil temperatures and atmospheric humidity. The permafrost can be anything from completely dry (devoid of water ice) to saturated with ice, filling, or even exceeding, the volume of the pore space of the soil.

In the present Martian climate, ice-rich permafrost should be abundant poleward of about 30 to 40 degrees latitude (encompassing the middle and high latitudes) in both the northern and southern hemispheres. In these regions the average subsurface temperatures are colder than the temperature at which the atmosphere saturates, a frigid 196 K (−77 C). Therefore water vapor in the atmosphere will migrate or diffuse into the soil and condense as ice in the pore space (the gaps between individual grains of soil). In this case we say that ground ice is stable with respect to the current climate.

In these stable regions ground ice will easily occupy the pore space within the soil, and may account for as much as half the volume of the soil. Relatively high summer surface temperatures will keep this ice from occurring too close to the surface. Instead, it will be buried beneath a protective blanket of dry soil somewhere between a few tens of centimeters to more than a meter deep where average subsurface temperatures stay below 196 K.

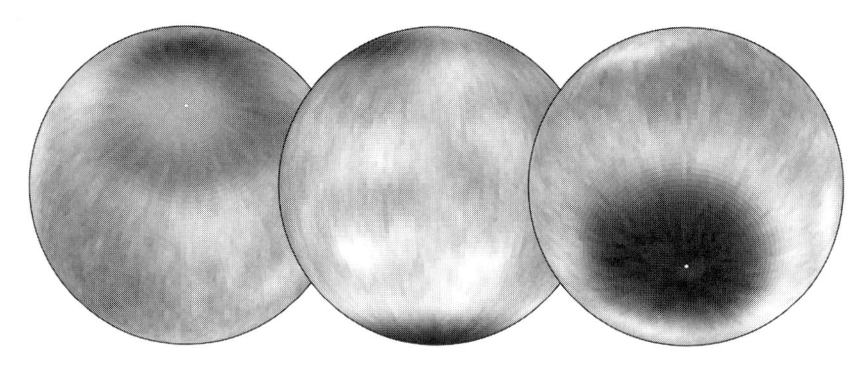

Figure 4.8 Epithermal, medium-energy neutrons measured by the Mars Odyssey spacecraft. Lower neutron densities (shades of dark gray to black) indicate an abundance of subsurface hydrogen and suggest abundant water ice as the source. High neutron densities (shades of light gray to white) indicate a dearth of hydrogen and an absence of ice. Ground ice a meter or more below the surface cannot be detected. These three globes of Mars rotate from north to south and east to west (left to right). The northern hemisphere (left) shows a weak signature of ground ice, obscured by thick seasonal CO_2 frost. The unobscured southern polar region (right) exhibits a strong signature of ground ice. *Credit:* NASA, Mars Odyssey, map data courtesy of Dr Bill Feldman.

While theorists predict the existence of abundant ground ice in the Martian soil, the presence of such ice can be inferred by the detection of abundant hydrogen. Two instruments aboard the orbiting Mars Odyssey spacecraft, a gamma ray spectrometer and a neutron spectrometer, detected abundant hydrogen. Galactic cosmic rays normally bombarding Mars generate high energy neutrons and gamma rays in the soil. After interactions with the soil itself, these gamma rays and neutrons are eventually expelled from the surface to be detected by the orbiting spacecraft.

The soil chemistry in the upper meter of the Martian surface influences the energy spectrum of both neutrons and gamma rays. Hydrogen holds by far the greatest sway on each. For example, a profound absence of medium-energy epithermal neutrons signifies deposits rich in hydrogen. While hydrogen could be present as hydrated minerals, water vapor, or thin films, only ground ice exhibits such a dearth of neutrons. For example, Figure 4.8 maps the quantity of epithermal neutrons expelled from the surface. In the southern hemisphere few epithermal neutrons are observed poleward of 60 degrees latitude, indicating that abundant ice persists only a few centimeters beneath the surface, happily in agreement with the predictions of the theorists.

Just equatorward of 60 degrees, ground ice likely resides at greater depths, beyond the view of the Mars Odyssey instruments. In the equatorial regions slight variations in the neutron density reveal something of the soil chemistry and the degree of hydration of minerals, but an absence of ground ice is also evident. In contrast to the southern zone, the northern regions of Mars present a complex neutron signature. During the season when these data were collected, seasonal CO_2 frost blanketed the surface, reaching several meters depth at the north pole. This thick seasonal dry ice frost obscured the detection of hydrogen and ice in the soil, though some ice is evident near the periphery of this seasonal frost layer.

Buried ice from the past

Large masses of nearly pure, buried ice occur commonly on Earth within the permafrost. These deposits can range in size from less than a centimeter to tens of meters in thickness. Some deposits may be buried remnants of glaciers or frozen lakes, while others may have formed in place as the ground froze by drawing in ground water from nearby soil, lakes, or rivers into the freezing zone.

On Mars, similar deposits of nearly pure ice may exist. The Martian geology presents abundant evidence that water once flowed on the surface: valley networks, outflow channels, and even shorelines and lake deposits in some low lying regions are seen in images acquired from orbit. If water from melted ground ice, hot springs, or even precipitation carved these channels, where is that water today? One possibility is that some of it remains in the form of frozen lakes covered by a thin, protective, and unfortunately obscuring veneer of soil a few meters thick. When we study images of Mars we could be looking at vast deposits of shallowly buried ice without realizing it.

Similarly, if Mars had a warmer climate in the distant past, as it cooled and froze to become the Mars of today, large masses of ground ice may have formed in place. Ice lenses, horizontal thin layers of pure ice less than a centimeter thick, are the most common form of buried ice on Earth and result from the freezing of wet soil. A pingo is a large mass of ground ice that forms when water from a nearby lake or stream is drawn through the ground to the bottom of a freezing ice layer. A mound results, 10 meters or more thick and several hundred meters in diameter. Ice lenses and pingos have not been identified in Martian images, and an astronaut may have to drill a core sample to verify their presence.

The story in the Martian equatorial region is somewhat different. Even though the temperatures are still way below freezing, the humidity is also extremely low. Any ice would rapidly sublimate and the vapor would be lost into the atmosphere despite any protective blanket of soil. This process in itself might leave behind collapsed pits or scarps, known as thermokarst, common in terrestrial permafrost from melting of ice-rich soil. Equatorial ice could only persist if it were trapped in the Martian surface beneath a layer of impervious rock.

Running water from frozen ground

The Martian landscape shows diverse geologic evidence that ice has influenced its genesis and evolution. Many of the giant outflow channels (Fig. 4.9) originate from collapsed and jumbled (chaotic) terrain. These channels, hundreds of kilometers in size, appear to have been carved out of the Martian landscape by vast floods of water in the past. They may have resulted from geothermal heat melting and thinning ground ice deposits. Geothermal heat can melt ground ice and circulate existing ground water, producing large pockets of subsurface water. Pockets of groundwater on a regional slope or hillside would be held in place by ice-cemented ground at the surface, a permafrost dam.

As the ice-cemented ground thinned or the hydrostatic fluid pressure of the groundwater increased, failure of this natural barrier would have released a flood of water across the Martian landscape. Ground failure could have occurred spontaneously. Conversely, it could have been triggered by a disrupting event. An ill-placed meteorite impact or marsquake could have compromised the permafrost seal and released a torrent of water, mud, and rocky debris across the landscape.

While the large outflow channels occurred in the Martian past, tantalizing evidence exists for liquid water on the surface of Mars today. Found at middle and high latitudes, small fresh gullies emerge from the tops of many slopes. The presence of incised channels only a few meters wide and a few hundred meters long suggests that water recently seeped from the ground and flowed a short distance downhill (Fig. 4.10). However, the surface of Mars today consists of permafrost kilometers deep, presenting something of a puzzle given the relative trickles of water that have been observed. Where does this water come from? Seasonal melting of ice-rich ground and fracturing of the permafrost by hydrothermal or freezing pressures in an aquifer may provide an explanation. Some scientists have even suggested that water may not be the erosive agent, but that CO_2 ice buried below the surface may melt periodically. The answer will require more study.

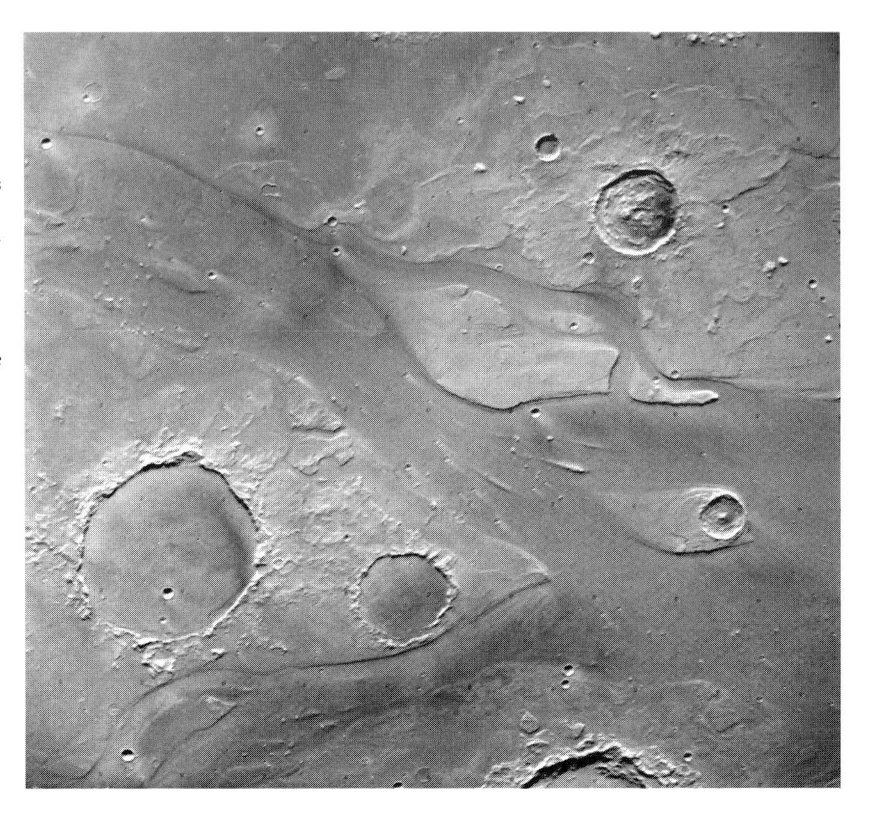

Figure 4.9 Catastrophic flood channels. Teardrop shaped islands and anastomosing channels indicate widespread erosion by liquid water. The formation of these channels does not require a warmer climate, since the abrupt release of water and inundation of the Martian surface can carve deep into the Martian landscape more rapidly than the water can freeze. Image width 278 kilometers.
Credit: NASA, Viking Orbiter.

On some of the oldest terrain numerous valley networks (Fig. 4.11) hint at a land of abundant liquid water and rainfall runoff, but scientists have suggested alternative explanations, such as geothermal melting of snow or melting and sapping of ground ice. These geothermal theories are more consistent with the geographic distribution of the valley networks. Many of these features are located on the flanks of volcanoes or localized areas of larger escarpments that would be candidate regions for geothermal activity.

All these erosional landforms formed throughout Martian history with a shift from valley networks in the more distant past to outflow channels and most recently small gullies. This trend indicates a changing Mars. Early in its history the climate may have been warmer and more humid than today, allowing for snow to accumulate in areas where it cannot accumulate today. Indeed, some of the diverse landscape far from the poles may even be glacial in origin. (A climate warm enough

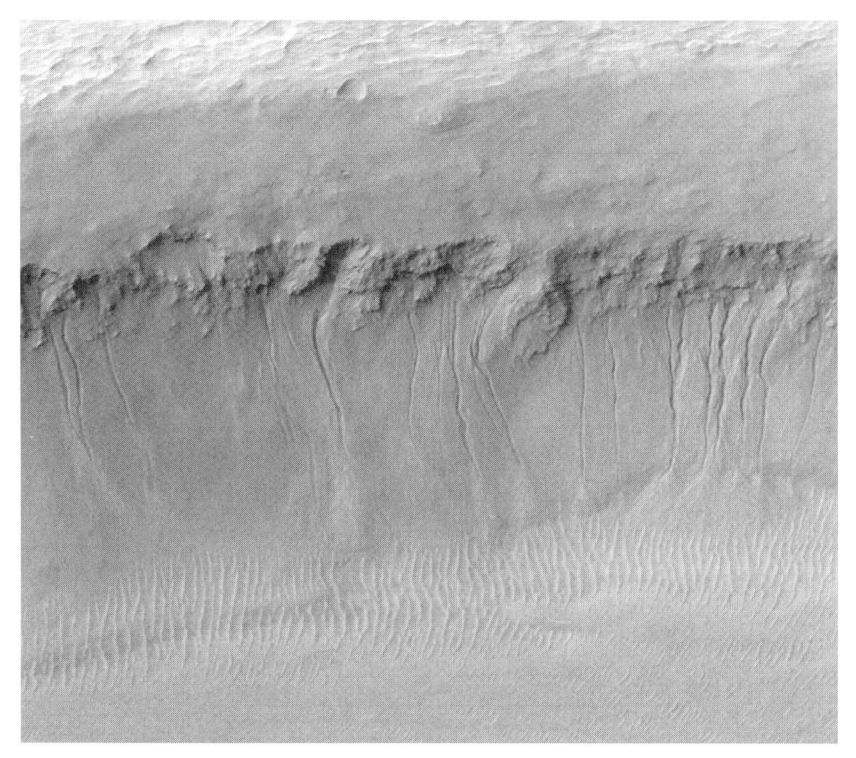

Figure 4.10 Recent gullies. Freshly incised channels a few meters across indicate a fluid eroded the soil to form these geologically young gullies. Water is the most likely candidate, but the source of liquid water on a globally frozen planet is a mystery. Ground ice melting within the source amphitheater at the top of the gullies or shallow groundwater beneath the permafrost have been suggested as water sources. Scene width is 2.1 kilometers. *Credit:* NASA, Mars Global Surveyor.

for rainfall may also have been possible; scientists are debating whether rainfall on Mars is necessary to explain observed geologic features.) An astronaut/time traveler standing on the surface of this younger Mars could have been faced with a wintry scene, with snow covered mountains, active volcanoes, and even hot springs. Ice-covered streams seeping from the ground could slowly carve the valleys we see today.

Later, as the climate cooled and dried, snowfall would become restricted to more poleward regions, and liquid water would become "cold trapped" in the permafrost. Geothermal heat, perhaps aided by circulating ground water, could then build underground reservoirs of melted ice. The scene for our astronaut/time traveler standing on the surface is now one of a cold dry Mars. While admiring the landscape, a nearby hillside would erupt in a landslide of rock, ice, and steaming water. As the subsurface reservoir drained, the land behind the source vent would collapse, no longer supported by the wet sediments. The flood of muddy water would form an icy crust as it cooled over a period

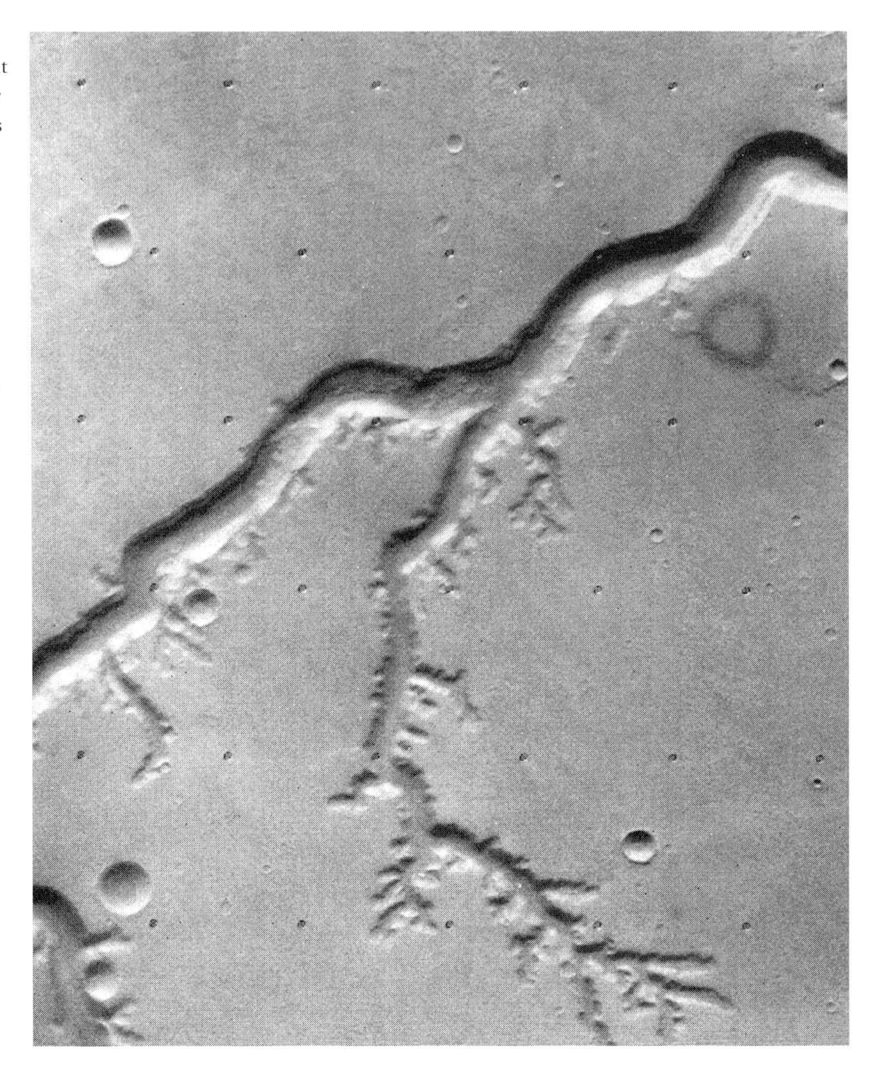

Figure 4.11 A water-carved Martian valley. Many short, blunt tributaries feed the larger valley (about 3.5 kilometers wide). This morphology suggests that the source of water emerged from beneath the surface. Erosion of the valley headwall may have occurred as seeping water undermined the surface, a process known as sapping. The source of water may have been melted ground ice. The absence of ever-smaller tributaries indicates rainfall or surface snow melt were not involved. *Credit:* NASA, Viking Orbiter.

of days, which would serve to protect the flowing water beneath. Such a flood might take weeks or even months to run its course as it traversed hundreds or even thousands of kilometers over the Martian surface.

Moving ice

Considerably less dramatic but no less important, at the higher latitudes in both hemispheres, the topography of crater rims, scarps, and hills currently exhibits a subdued and rounded character. Similarly,

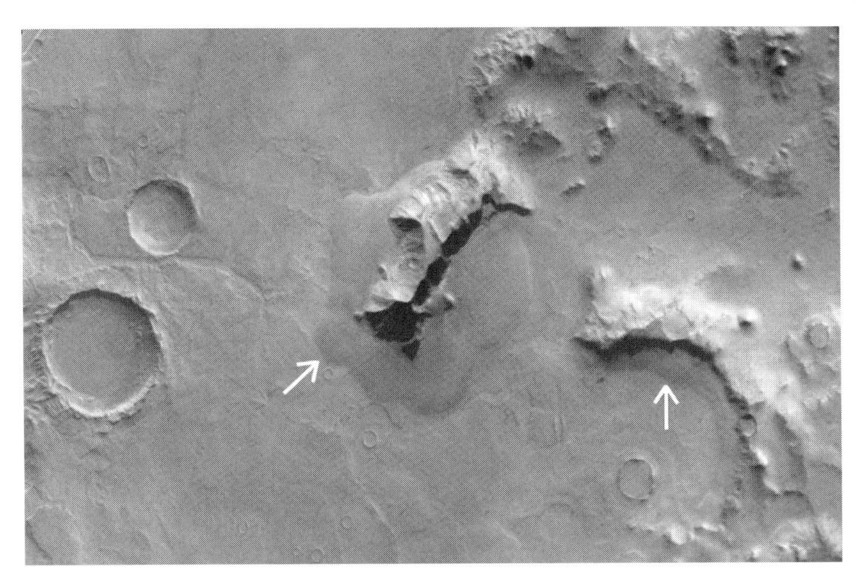

Figure 4.12 Martian rock glaciers? Rounded aprons (arrows) at the base of mountains and massifs suggest slow outward flow (in places as far as 25 kilometers) of ice-cemented rocky debris. Similar morphology, lobate shape and steep termini, are found on terrestrial rock glaciers, indicating a similar genesis. Rock glaciers creep along at a few centimeters per year, and so take many thousands of years to form. *Credit:* NASA, Viking Orbiter.

aprons of debris with abrupt and rounded, lobate termini surround many mountain knobs (Fig. 4.12). Scientists believe this appearance to be caused by downhill viscous movement of ice-laden permafrost.

In many mountain regions on Earth talus from eroding cliffs can become entrained with ice and snow, creating what is known as a rock glacier. This water ice acts as both a cement and a lubricant. Under the weight of the icy talus upslope, the ice deforms just as a glacier does and allows the rock and debris to glide past each other as a sort of slow motion fluid. At the same time, where a true talus slope would grade gently into the valley floor, the ice acts as a cement, holding the terminus of the debris at a steep angle.

Martian landforms appear similar in shape to terrestrial rock glaciers, and, if they indeed formed through the same processes, indicate the presence of subsurface ice. Additionally, since rock glaciers can take thousands or even millions of years to develop, they tell us that the subsurface ice in these regions has been active for a long time.

At the limits of vision

On Earth, ice more commonly influences geology on a small scale, creating polygonal fracture patterns, sorting small stones from large stones, and generating lobes of soil on hillsides, all a few meters or less in size.

Figure 4.13 Troughs on the Martian surface. This trough (indicated by arrows) is about 1 meter wide and 10 centimeters deep and is part of a network of semi-connected troughs around the Viking Lander 2. Similar troughs are found in the Arctic and Antarctic regions of Earth. They indicate subsurface fractures in ice-cemented permafrost. Their presence on Mars suggests that abundant ground ice occurs just below the surface.
Credit: NASA, Viking Lander.

These periglacial features are ubiquitous in terrestrial permafrost. We expect Martian permafrost to express itself in similar ways. With the highest-resolution images of Mars available from spacecraft, we are able to see only the largest of these features.

Polygonal patterns are the most common in the Earth's frozen wastelands. Ice-cemented ground undergoes contraction in the winter as seasonal temperatures fall. Fractures rupture across the ground as a result of thermal stress generated by the contraction process. Randomly oriented cracks coalesce to form a honeycomb network, typically 10 meters or more in diameter. When surface material such as dry soil falls into the crack, a depression or trough develops about a meter wide at the surface. The current climate conditions on Mars are ripe for polygonal fractures to form wherever ice-cemented permafrost is found.

One of the two Viking Landers that imaged the Martian scenery in the late 1970s touched down in a region where ground ice may be abundant below the surface. Intriguing shallow troughs, about 10 centimeters deep and roughly a meter wide, were observed around the landing site (Fig. 4.13). When mapped, these troughs composed fragments of

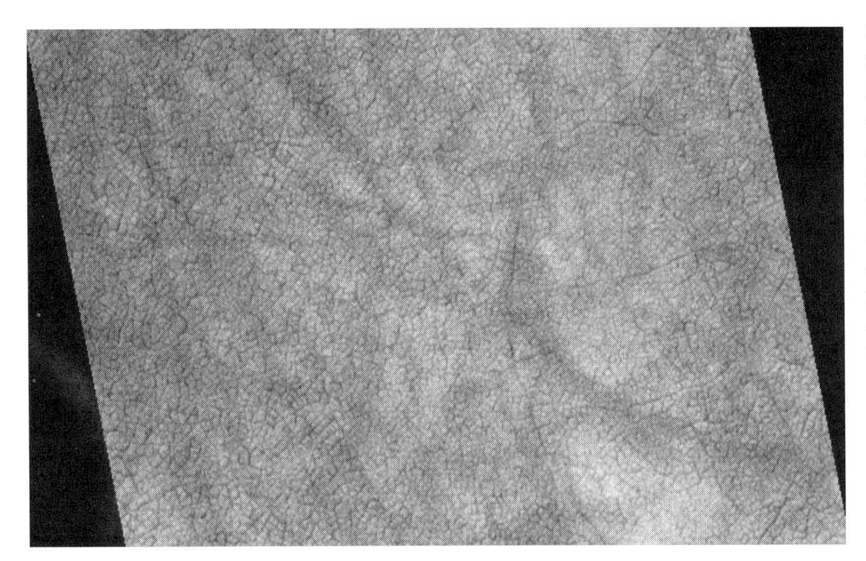

Figure 4.14 Polygonal patterns in the southern hemisphere. A honeycomb network of troughs, each about 10 meters in diameter, suggests that ice-cemented permafrost occurs below the surface over large areas. Fracturing of the ground during winter is commonplace in icy permafrost. Scene width is about 900 meters.
Credit: NASA, Mars Global Surveyor.

a polygonal pattern about 10 meters in diameter. One interpretation suggests these troughs are the surface manifestation of fractures in the ice-cemented ground below the surface.

The Viking Lander provided us with the first glimpse of potential periglacial features on Mars. High-resolution images from a more recent spacecraft, Mars Global Surveyor, regularly presented polygonal patterns similar in size and shape to those of Earth (Fig. 4.14). These images begin to paint a picture of abundant ground ice and a landscape where ground ice exerts its influence. Other smaller features await future missions with improved imaging abilities or perhaps direct exploration by astronauts.

Impact craters in the permafrost?

A class of impact craters on Mars (termed rampart craters), with rounded ejecta deposits (material thrown from the forming crater) and debris bunched up at the outermost reach, give the appearance of a meteor striking mud (Fig. 4.15). They differ from normal impact craters where ejecta is strewn relatively evenly outward from ground zero. A leading hypothesis contends that these craters formed when ground ice melted or vaporized from the heat of the impact and were

Figure 4.15 A rampart crater. Many impact craters on Mars, such as this 18 kilometer diameter crater, have lobate ejecta radiating from the crater rim giving the appearance of a meteor striking mud. The prevailing view is that these craters formed in ice-rich permafrost, where melted and vaporized water ice became entrained in the shower of debris ejected from the collision. *Credit:* NASA, Viking Orbiter.

entrained in the ensuing shower of debris. The water and water vapor would act to lubricate the ejecta giving it this characteristic lobate appearance.

Such craters are commonplace on Mars and occur everywhere on the planet, but only the largest craters near the equator exhibit rampart ejecta. As we move poleward many smaller craters manifest this characteristic ejecta. This strong correlation with geographic location is explained by the depth of excavation of an impacting body. Larger craters are formed by larger objects that excavate to greater depths. At the equator, where ground ice is unstable and can sublimate into the atmosphere, impactors need to reach great depths in the subsurface

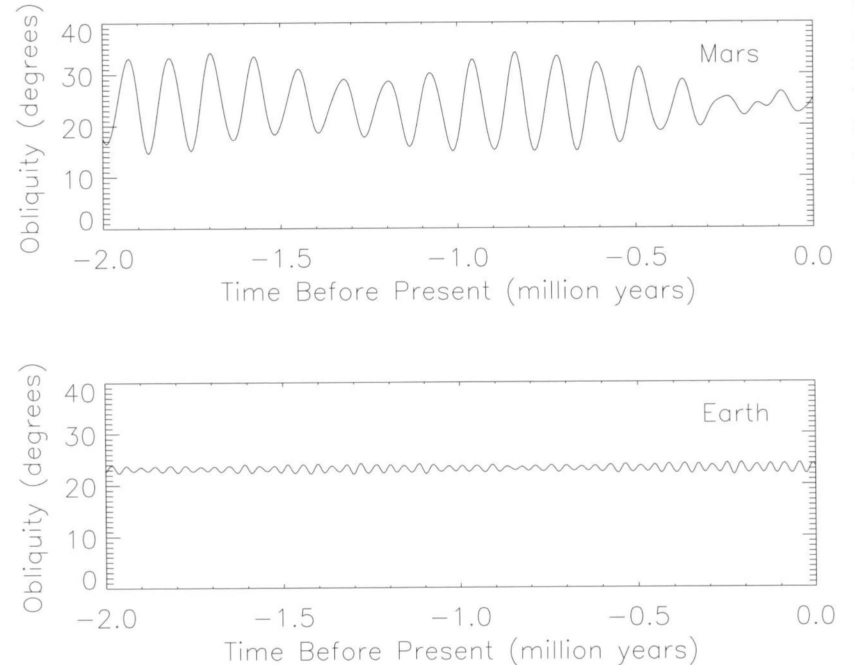

Figure 4.16 A comparison of the obliquity histories of Mars and Earth. The obliquity, the tilt of the planet's spin axis relative to its orbital plane, controls the seasons. This tilt undergoes changes induced by gravitational interactions with the Sun and other planets. Shown here are the histories of obliquity for the past 2 million years. Small changes for Earth are known to have a dramatic effect on Earth's climate, enough to cause ice ages. The influence of large oscillations of the Martian obliquity will be even more dramatic.
Credit: M. Mellon.

to locate water, possibly deep enough to reach unfrozen ground water. Closer to the poles, where ground ice is stable, rampart craters form more easily from small objects excavating to shallow depths.

Climate change

Small oscillations in Earth's orbit produce climate changes dramatic enough to cause ice ages. These oscillations result in shifts in the distribution of solar energy on the Earth's surface. For example, an increase in the tilt of the polar axis, the obliquity, can result in more sunlight striking the polar regions in the summer, producing warmer temperatures. A decrease in the obliquity will lessen the amount of solar energy reaching the surface, and summer temperatures will be lower.

On Mars, such oscillations are considerably larger, partially as a result of the red planet's closer proximity to the gravitational influences of Jupiter and the absence of a large stabilizing moon (Fig. 4.16). Therefore, we expect changes in the Martian climate to be more impressive. Currently, the Martian obliquity is 25.2 degrees but oscillates by more than 10 degrees away from this value. For comparison, the Earth's obliquity stays within about 1.5 degrees of its current value.

An increase of just a few degrees in the obliquity will cause a dramatic change in the stability of ground ice in the permafrost. At higher obliquities the Martian polar regions will receive more sunlight causing higher surface temperatures, rapid sublimation of water ice, and increased humidity in the atmosphere. With more water in the atmosphere, the temperature at which the atmosphere saturates will rise, to perhaps a relatively toasty 220 K (−53 C) or higher. This excess water in the atmosphere will migrate into the soil and condense in regions where the average soil temperatures are less than this new higher saturation temperature. Such regions would include the equator.

Likewise, a decrease in the obliquity would reduce the amount of sunlight incident on the polar surface. The resulting colder polar surface temperatures would trap water vapor from the atmosphere, effectively drying it out. Ground ice that had been stable in the equatorial permafrost when the obliquity was higher would experience a change in climatic conditions – a dry atmosphere. In these circumstances ground ice would sublimate and be carried away in the desert wind, but would eventually find its way back to the polar deposits.

Oscillations of this nature occur on Mars about every 100,000 years, a relatively short period of time on geologic scales. Similar oscillations occur in the eccentricity (a measure of how circular the orbit is) and the time of perihelion (season of closest approach to the Sun). However, these orbital properties are less important than the obliquity in controlling the climate. It is periodic changes in the Martian orbit like these that are believed to be responsible for the polar layered deposits, although the precise mechanism for accumulating the layers remains a mystery. Interestingly, estimates of the total amount of water ice involved in this migration from the polar deposits to the permafrost and back to the poles in one obliquity cycle is enough to account for one layer in the polar layered terrain.

Other deposits of ice in non-polar regions may be present today as a relic of a past climate; similar deposits of relic ice are found at some locations in terrestrial permafrost left over from the last ice age. On Mars, ground ice left by a recent past climate may be present at latitudes where it is unstable today. Knowing where the ice is and how it got there can provide important clues to Martian climate history.

An elusive resource

Human habitation of Mars is a definite possibility in the future. Colonists would need many resources to survive, but water would be one of the most vital requisites. Although to some degree recyclable, settlers would need water in abundance not just for human consumption but for cultivating plants and raising animals. Water might also be used as a combustible fuel by separating hydrogen from oxygen, and more ambitious settlements might use water for industrial projects, such as separating mineral ores and chemical processing. To ferry water from Earth would be a costly and difficult undertaking. Therefore, locating adequate and accessible supplies of water on Mars is desirable.

Locating human settlements near the most abundant and obvious sources of surface water, the polar deposits, has the appearance of an easy answer. Certainly, water is abundant; the north polar deposit alone is estimated to contain more than one billion, billion liters of water. Tapping into the reservoir would not be difficult, since it would contain only a small fraction of dust and sand. Melting and filtering could remove any solid contaminants. Salts dissolved from soil grains or brought in with windblown dust comprise a chemical contaminant that could be easily removed by distillation.

Living in the Martian polar regions presents other problems, however. Extreme cold, even by Martian standards, would test the limits of machinery and people. Even in the polar regions of Earth, where temperatures are considerably warmer that at the poles of Mars, the harsh environment represents a life-threatening hazard. Reduced sunlight would make it difficult to grow plants, particularly during the polar night. More importantly, solar power would not be an option. While the Martian day is just under 25 hours long, the winter pole remains in darkness for more than 11 months. An alternate power source would be required to provide adequate heat, to power day-to-day activities, and to run the water purifying equipment. And of course, winter deposition of CO_2 frost could bring outdoor activities to a standstill.

The equatorial regions would present a much more favorable environment for habitation. At times, settlers would experience even Earth-like temperatures. Typical temperatures, though at times still harshly cold, would be manageable. Abundant direct sunlight would be available for food production and for powering life support systems (heat, water, and air) and industrial projects.

Water resources at the equator would be limited and elusive. In the present climate the equatorial region is too dry for ice to persist at or near the surface. Ice could exist only in trapped reservoirs beneath impervious rock strata, unable to sublimate and migrate to the surface where it would be lost into the atmosphere. Finding such a deposit, if one exists, would be difficult and would require expert knowledge of the local geology that would come only from long-term study by resident geologists.

Drilling a well kilometers deep to reach below the permafrost in search of ground water presents problems of its own. It is not known if ground water even exists beneath the permafrost, let alone where. Current theories suggest that if enough water exists on Mars to fully saturate all the pore space of the cryosphere (the global permafrost), any excess water would settle in the lowest reaches of the crust as ground water. How much water exists on Mars? Is there enough to provide for groundwater or is it all locked up as ice in the permafrost? Presently, we do not know. Furthermore, drilling deep wells in the Earth's permafrost introduces many engineering difficulties. Add the limited resources available to solve engineering complexities and drilling deep wells on Mars could present insurmountable problems.

An alternative may be to extract water from the atmosphere by condensing the water vapor transported by the Martian wind. Since this water is ultimately provided by sublimation from the polar deposits, this method would let nature do most of the work to import water from the polar regions. The dry Martian atmosphere, however, does not contain a great deal of water for extraction, so atmospheric water would only provide a limited supply.

The middle latitudes (roughly between 30 degrees and 60 degrees) present a more likely choice. Although ground ice would be integrally mixed with the soil, it could be easily extracted from blocks of icy soil by melting and filtering. Locating purer ice in a frozen lake or other massive ice deposit would be an added bonus, but not necessarily a requirement. Aside from ground ice at middle latitudes, the presence of small, geologically young gullies suggests that in some regions liquid water may occur fairly close to the surface, perhaps within a few hundred meters. Although drilling may be necessary to tap into this potential

Figure 4.17 Winter water frost at the Viking Lander 2 site. This ice condensed from excess atmospheric moisture on the cold ground surface. CO_2 frost can also condense at middle and high latitudes, hampering outdoor activities near a human settlement. However, winter CO_2 frosts are thin and often occur only at night at these middle latitudes.
Credit: NASA/Viking Lander.

reservoir, the depths may be more practical with today's technologies, and extracting liquid water from the subsurface would be easier than mining for ice. In addition to water, the middle latitudes would also offer more sunlight and warmer temperatures than the polar regions and an absence of thick winter CO_2 frost deposits.

Hazards of living on ice

Life on Martian permafrost presents new hazards. Heat leaking into the ground from shelters could melt ground ice, causing settlers and their equipment to sink into Martian mud. If melted water seeped deeper into the soil, local collapse of the surface might follow. Even a small amount of heat leakage into ice-rich permafrost could raise the soil temperature and cause sublimation and local collapse. Either way, the structural integrity of the human settlement's living space would be compromised, and on Mars where the atmosphere is unbreathable, this could be a fatal problem.

Placing surface structures on stable bedrock or on elevated pilings above the ice-rich permafrost would be one solution. Human habitats would also need to be shielded from space-borne radiation in the absence of a thick atmosphere and planetary magnetic field. Shelters could be built below ground in ice-free permafrost, if an ice-free area is available. Another solution might be to refrigerate the ground to keep it frozen, a technique used in the construction of the Trans-Alaskan Pipeline. Such problems would need to be solved well ahead of time, requiring more complete knowledge of the distribution and nature of ground ice and general properties of the Martian permafrost.

Life?

Water is essential for life as we know it. Mars has revealed a myriad water-carved channels and valleys where life could have emerged when climate conditions were warmer than they are today. Today, Mars is frozen and life would have to adapt or die. To adapt, it would have to develop the ability to lie dormant for long periods when frozen conditions prevent growth. In Antarctica, cryptoendolithic lichens survive just beneath the surface of porous rocks, where they are protected from the cold. Sunlight can warm them to temperatures above freezing during the brief summer and provide liquid water from melting snow. These lichens otherwise lie dormant and frozen during the remainder of the year. Martian life may have to wait considerably longer.

Ground ice containing dormant microorganisms could be heated periodically by geothermal activity or dramatic changes in the climate. Local volcanic vents could warm the surrounding soil and melt the ice. Martian life could metabolize and live off chemical compounds produced by geothermal energy in much the same way terrestrial life persists near sea floor hydrothermal vents. When volcanic activity wanes, microbes could again become dormant as they are trapped in the freezing sediments, awaiting the next eruption.

Similarly, during changes in the Martian climate induced by orbital oscillations, greenhouse gases might be released into the atmosphere (water vapor and CO_2 are likely candidates). These gases would warm the planet's surface by allowing sunlight to reach the surface, while at the same time absorbing thermal energy that would otherwise escape to space. Greenhouse gases may in turn warm the planet sufficiently that some small regions of the surface can support liquid water. Organisms might thrive in these little oases until the orbit and climate revert to unfavorable conditions.

Portions of the Siberian permafrost, undisturbed for 3 million years, have been found to contain viable bacteria, protected throughout their long wait, while sealed in ice at low temperatures. If periodic Martian warming was driven by obliquity cycles, a Martian organism perhaps would only have to wait for hundreds of thousands of years.

If, on the other hand, Martian life could not adapt to the general freezing of its planet, it would have died. Scientifically though, all is not lost. Dead organisms and organic material might be trapped in the permafrost in places where the ice is sedentary, undisturbed for several billion years. Although now unrevivable, the organisms would be well preserved within the ice, providing biologists with a scientific gold mine. The discovery of the preserved organic remains of Martian life in ancient permafrost would provide clues toward the origins of life on Earth and the possibility of life elsewhere in the universe.

The future

Although Percival Lowell's view of Mars may have been clouded by the excitement of recognizing that Mars was an Earth-like terrestrial planet, a frozen Mars remains an exciting place for new icy discoveries. Ice is centrally important to many aspects of the Martian environment. It affects the climate, sculpts the landscape, and is an essential resource for life originating on Earth or Mars. Ice is also rather dynamic, migrating from pole to pole, or between the poles and the global surface layer of soil, in response to changes in the climate. Much scientific research focuses on how ice affects Mars, and many questions remain unanswered. How much water exists on Mars? How long has Mars been frozen? Where is all the water today and is it accessible at the Martian surface or permanently locked away in deep underground permafrost? How do we locate water and ice? Some of these questions must be answered before humans can set foot on the Martian surface. Human exploration will be needed to complete our understanding of the Martian landscape.

PAUL M. SCHENK
Lunar and Planetary Institute in Houston, Texas

The ice moons of Sol

Our home planet is often considered a "Waterworld," but we need to go beyond the asteroid belt, into the realm of Jupiter and beyond, to find worlds made mostly of water and ice. Human cognizance of the outer solar system has been slow to reach maturity and naturally breaks into three distinct eras: the telescopic era from 1610 to 1979, the Voyager era from 1979 to 1996, and the Galileo–Cassini era from 1996 on. Each era has brought its own revolution.

The telescopic era began when Galileo pointed an optical device skyward, shook Christendom, and forced us to recognize that Earth was not, after all, the center of all things. After Galileo, most observers were concerned with establishing the basic framework of the outer solar system, including counting all the moons and understanding the basic nature of Saturn's rings (which proved to be a disk of icy dust, pebbles and boulders). By the twentieth century, the outer solar system had become the slum of the astronomical world. Grindingly dull photometry and polarimetry studies were the norm. Although spectroscopic studies of the composition of the planets began in the 1940s, it has been only in the past 30 or so years that we began to understand how water-rich the outer solar system is.

As the space age dawned in the 1960s, the outer solar system seemed a vast, empty, and static place. The placid orbital dance of the giant gaseous outer planets and their cold satellites was disturbed only by stray comets or by occasional storms in the chilly clouded atmospheres. Indeed, one justification for exploring these worlds was the hope that they might contain pristine remnants of the birth of the solar system. That hope is nearly (but not quite) gone as the outer solar system has proved to be much more dynamic in ways we could not have guessed 40 years ago.

With the arrival of the Voyager spacecraft at Jupiter, the year 1979 brought the discovery of active volcanoes on Io and the intensely cracked and fractured surface of Europa, and with it a realization that our expectations could no longer be any guide to what we might see next (Fig. 5.1). Subsequent years saw a blizzard of discoveries, from tiny resurfaced moons to volcanoes (and geysers) at the edge of the solar

Figure 5.1 The diverse moons of the solar system. This mosaic of Voyager images shows the moons with sizes and relative brightness preserved. Each planet is presented in rows. From top to bottom: Earth's Moon, Jupiter satellites (including the four large galilean satellites), saturnian satellites (including Titan at center), the uranian satellites, Neptune satellites (including large Triton), and Pluto's moon Charon. *Credit:* Timothy Parker and Paul Schenk/LPI.

system, to a family of stray comets buzzing near Jupiter, some of which are pulled in and collide with Jupiter or one of its satellites. Indeed, even the orbits of the gas giant planets have changed with time. We have also seen the discovery of the trans-Neptunian Kuiper–Edgeworth Belt, pushing the edge of the known solar system out much farther than previously thought. This wide belt is crowded with hundreds and perhaps thousands of small, icy, asteroid-like bodies, remnants perhaps of the formation of the solar system.

With the construction and launch of advanced robot explorers on dedicated missions to these outer planets, we have witnessed yet another revolution, starting with Galileo's arrival at Jupiter in late 1995. With high-resolution images and spectral maps of the surfaces of the jovian moons, and gravitation and magnetic probing of their interiors for the first time, we see moons vastly more complex than even Voyager's discoveries led us to believe. One of the more subtle but important discoveries has been the geologic, geophysical, and geochemical uniqueness of each of these worlds (Fig. 5.1). Many of these icy bodies must be considered planets in all but name.

We are still at the dawn of the Space Age, and perhaps the most profound discovery of this time has been that the liquid water layers and complex atmospheres on some of the ice-rich satellites of the outer planets may support organic chemistry or perhaps even primitive life. Whether such life actually exists remains speculative. Life of course requires more than organic chemistry. Energy and food (in the form of carbon, nitrogen, and oxygen) are also present on these satellites, although perhaps not always in the right form at the right time. In some cases, all that might be needed is a little heat to start some interesting reactions. Although Mars remains the center of speculation in this regard, the possibility of environments conducive to the formation of life or its basic elements in the outer solar system cannot be ignored. Whether this new speculation will hold up after another 40 years remains to be seen.

Moon madness

Before we examine some of these moons in detail, let us look at the satellites collectively and the processes that shape them. The satellites of the outer planets can be divided into three general categories based on size (Fig. 5.1). The smallest are sometimes referred to as "rocks." These are usually less than 200 kilometers across and are typically found either very close to or very far from their parent planet. Little is known about these lumpy bodies but most are thought to be left over from planetary formation or are captured asteroids or comets. This chapter will focus on the remaining satellites, the middle-sized moons and the large, planet-sized moons.

There are 15 major satellites between 400 and 1,000 kilometers in size. These so-called middle-sized moons are found mostly around Saturn and Uranus, although Neptune is suspected of once having had a family of such bodies. The satellites that attract most of our attention are the six large planet-sized moons. These range in size from 2,700 to 5,260 kilometers in diameter. Two are larger than Mercury and all are larger than Pluto. Four orbit Jupiter, forming the quartet Io, Europa, Ganymede, and Callisto known as the galilean satellites, after their discoverer. The other two, Titan and Triton, orbit Saturn and Neptune respectively.

Water! Water!

What motivates continued exploration of the outer solar system more than any other single factor is water. The outer solar system is awash in water, most of it in frozen form. Fifteen thousand years ago, most of Canada, Scandinavia, and large parts of the United States were blanketed beneath several kilometers of ice. We should consider ourselves fortunate. The entire surface of Europa is completely smothered by a water layer at least 100 kilometers thick, the outer portions of which are frozen as ice. Telescopic spectra show that the surfaces of most of the satellites orbiting the outer planets (and Pluto, too) are also covered by water and other ices. Only Io lacks water ice, although the frozen volcanic sulfur dioxide on its surface technically qualifies it too as an icy moon.

The usefulness of water for sustaining living organisms depends on where and in what form it resides on an icy satellite. Solar system ices are much less dense than common rocks, and have a strong tendency to rise buoyantly towards the surface and be concentrated in the thick outer layers. Ice therefore replaces the more familiar silicate rocks as the bedrock and crust-forming substance on these satellites. We are not accustomed to visualizing ices or planetary surfaces in this way, but it turns out that most of the processes responsible for shaping the surface of the Moon and other planets operate in much the same way on these satellites. This includes the melting of crustal and mantle material and extruding it on the surface, a process known as volcanism regardless of whether water or molten silicate lava is involved.

What we do know about the composition of the icy satellites has not been known for very long. Theoretical work by John Lewis in the early 1970s and others suggested that water ice would be very abundant in the outer solar system. Most ices are bright and relatively colorless at visible wavelengths, but have prominent and diagnostic spectroscopic absorption bands in the near- and mid-infrared spectrum. In the early 1970s, telescopic spectroscopic observations confirmed the presence of abundant water ice on the surfaces of three of the four large galilean satellites of Jupiter. Although no water ice has been detected on volcanically active Io, SO_2 frost has been identified there as well. By 1984, water ice had been detected on all the major saturnian and uranian satellites as well.

How far into the interiors of these moons do these ices extend? As a rule, each of the icy moons has a low bulk density (generally equal to or up to two times that of ice). Once this was established in the 1970s and 1980s, it became clear that these satellites are composed mostly of water (and other) ices. In most cases, between 30% and 70% by mass of these satellites is composed of water ices, forming outer shells hundreds of kilometers thick. Kilometers-thick layers of water ice on a planetary body have obvious value as a source of liquid water and oxygen for anyone attempting to work or survive there. Whether this water could supply a thirsty Earth in the distant future is unclear, but there is enough water ice on the largest satellite, Ganymede, to nearly equal the mass of Earth's Moon.

Organic stews?

With increasing distance from the Sun, additional ices which formed at even lower temperatures were expected to be stable in the solar nebula. Models of the nebula suggested that more exotic ices such as methane (natural gas) and ammonia would also be stable and form at least 10% of some of the outer satellites. Methane, in combination with oxygen, makes a useful energy source if enough can be economically concentrated. These estimates were based on considerations of low-temperature condensation processes within a nebula of roughly solar composition. Unfortunately, we do not yet have a firm grasp on the amount of these and other volatile ices present on the outer satellites.

Regardless of their economic potential, the satellites are vital scientific outposts. Their ices are key indicators of the conditions and composition of the solar nebula, the nebula that gave birth to the Earth. These include carbon monoxide, methane, and nitrogen, as well as simple carbon compounds such as methanol and formaldehyde, depending on the conditions of condensation. Indeed, some of these species have been discovered in comets, seemingly confirming these conclusions. Comets originate from belts of comets and other debris orbiting beyond Neptune. It seems likely that these small objects, some only a few kilometers across, have escaped the deformation and melting experienced by many of the planetary satellites. Their composition may prove to be directly related to composition and conditions within the outer solar nebula.

Recently, traces of frozen oxygen have been identified on Ganymede and carbon dioxide on both Ganymede and Callisto, probable byproducts of the impact of jovian magnetospheric ions on to surface water ice. As instrument technology advances, we may detect the ammonia and methane ices predicted on the saturnian and uranian satellites. The first tentative identification of the much-anticipated but elusive ammonia ices has only recently been made on distant Charon, satellite of Pluto.

Similar in size to Ganymede and Callisto, Titan is unique in being the only satellite to possess a substantial atmosphere (the surface pressure on Titan is 50% greater than on Earth). Titan is currently shrouded in a bland orange methane haze 200 kilometers above the unseen surface. Although the atmosphere is dominantly nitrogen, the variety of hydrocarbons such as acetylene and hydrogen cyanide that have been identified so far make this organic soup more toxic than swamp gas. The presence of even some methane in the atmosphere is almost certainly a reflection of the composition of the interior. The interiors of Titan and the other saturnian satellites apparently include more than just water ice and rock.

Only when we reach distant Neptune and Pluto can we actually detect any of the more volatile ices in large abundance. Solid methane, carbon dioxide, carbon monoxide, nitrogen and water ices have been identified on Neptune's large moon Triton, and on Pluto itself. We can identify ices on the surfaces of these satellites, but identification of the non-ice component has proved very difficult. Ice and snow are highly reflective, as any mountain climber will attest. Despite definitive identification of water frost and ice, some of the satellites, however, are rather dark (Fig. 5.1). The uranian satellites reflect only 20–40% of incoming sunlight. Callisto, with a brightness of only 18%, is nearly as dark as the Moon (which appears bright to us only in contrast to the dark night sky). Ice is so transparent, however, that only a small amount of dark material needs to be mixed in to turn ice very dark (those who live in areas that experience significant winter snowfalls will recall the ubiquitous piles of "dirty" snow on the roadside several days after a snowstorm).

Clearly the surfaces of many of these satellites contain, or are "contaminated" by some dark but as yet unidentified "rocky" component. In the case of Callisto, ice may make up only 15–45% of the surface layer. Although we have not drilled on these satellites, it appears that the rocky material does not extend very deep into their crusts, which are most likely dominated by water ice. This is demonstrated by the bright ejecta patterns surrounding recent impact craters, which are evidence for the excavation of buried ice. Thus, this dark material may be merely a surface contaminant.

The Galileo infrared instrument, NIMS, has found what may be the spectral signature of organic compounds on Ganymede and Callisto. Although other interpretations of the data are possible, the spectral bands resemble those of tholins, residue formed when organic materials are subjected to electrical discharges. Hydrated or ammoniated silicates related to clays may also exist on both satellites. These minerals could be the products of alteration between surface ices and cometary materials produced during the heat of meteorite impact. Or they may also have been brought in by the slow steady rain of small meteorites, mostly derived from comets, which are known to be rich in carbon- and nitrogen-bearing species. Many satellites, including Ganymede and Callisto, are also subject to bombardment by ions and other charged particles contained within the planet's magnetic field. Such compositions make sense in light of the accumulation of meteoritic debris and Jupiter's radiation belts.

The geologically young surface of Europa has not had time to accumulate much debris from the rest of the solar system, but has dark reddish material nonetheless. The origin of this material appears to be different from that on Ganymede and Callisto. The composition of the dark material is in dispute; it could be either salts or sulfates. If salts, the most likely origin is extrusion or eruption from inside, possibly from a salty ocean beneath the icy crust (more on this later). If sulfates, this could be a byproduct of Io's sulfurous volcanic activity. Io's volcanoes are the chief source for protons within Jupiter's radiation belts, and this material sweeps out to Europa's orbit over time, where it can be implanted on the surface and react with the surface ice. The distribution of this dark material is consistent with some magnetospheric control on its formation, but it may be some time before we can be sure whether it is Europa's subsurface salty ocean or Io's sulfur we are observing.

Most of the other satellites of Saturn and Uranus have relatively dark surfaces, despite being ice rich. We should know more about the composition of the saturnian satellites after Cassini arrives in 2004. Perhaps they too have altered organic molecules and other meteoritic debris mixed into their soils. One exception must be Iapetus (Fig. 5.2). This two-faced moon has fascinated astronomers since the 1600s when Cassini noticed that it was much brighter on one side than the other. Voyager did not get a great view of Iapetus but saw enough to tell us that its surface resembles a baseball, with one of the two leather panels painted black. One side is bright and icy (as are the poles) and resembles icy, heavily cratered Rhea, which is its twin in size. The other side is very dark and nearly featureless. (Apparent lack of features could be a result of the short Voyager exposure times used.) This hemisphere is so dark it reflects only a few percent of the sunlight striking it. Spectra of this hemisphere are poor in quality but may be consistent with the presence of carbon-rich materials, including tholins, on the surface of this satellite, too.

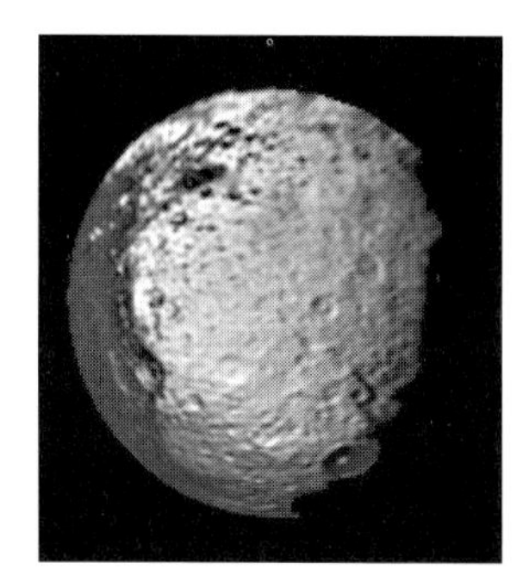

Figure 5.2 Two-faced Iapetus, satellite of Saturn. Voyager got only a glimpse of this enigmatic moon, with one side as bright as snow, the other as dark as charcoal, which has puzzled astronomers since the 1600s. Iapetus will be examined at close range by the Cassini spacecraft. *Credit:* Paul Schenk/LPI.

The chemical complexity of the icy satellites may be related to their locations. Water and ice are cosmologically abundant and their abundance in the outer solar system is no longer a surprise. The circumplanetary nebulae each satellite system formed in may have differed from the solar nebula at large. The large protoplanets may have radiated their own heat, altering the thermal and chemical equilibrium of the space around them. Hence, satellite composition may differ somewhat from that of comets, or Pluto and Triton, both of which originally formed in solar orbit (Triton was probably a solar-orbiting body captured by Neptune after formation). Titan may thus have more ammonia and methane than its cousins Ganymede and Callisto because of cooler conditions around Saturn.

These conclusions depend on our initial and somewhat uncertain assumptions. Carbon appears to be a key player in this scenario. The thermal and chemical state of the nebula will determine whether the carbon forms CO (sucking up oxygen and resulting in less water), or CH_4 (releasing more oxygen for water formation). The formation of ammonia is also dependent on these conditions. Working out the details of the chemical condensation process in a nebula that no longer exists is not a simple task, however. Thorough knowledge of the internal composition of each satellite system would be most useful in addressing these problems, as would mapping the composition of nebular disks surrounding other stars.

Energy to spare

Development of complex organic chemistry, and life itself, requires an energy source to sustain chemical and organic reactions. Geologic processes on the surfaces of planetary bodies are the best indicators of whether a planetary body has internal heat and therefore energy. The most obvious geologic process is the melting of crustal and mantle material and its extrusion on the surface, a process known as volcanism regardless of whether ice or rock is involved.

Although the outer solar system appears to be locked in a permanent ice age, the bodies populating this region of space are surprisingly complex geologically. Some are completely battered by impact craters and have been quiet for eons. Others, despite frigid surface temperatures of −145 C or less, have undergone intense volcanic activity in the more recent past. Geologic activity is ongoing on at least two satellites, Io and Triton, and possibly several others, including Europa and tiny Enceladus. Still others, such as Ganymede, Dione, and Miranda, have been intensely disrupted by faulting and volcanism in the more distant past, although they are inactive today. Before we examine each world in detail, let us look at how volcanism can be sustained on worlds where the surface temperature never exceeds 150 K, even in summer.

A major problem presented by the variety and extent of volcanic and tectonic resurfacing on the icy satellites is to identify heat sources powerful enough to melt so much ice, to force these lavas to the surface, and to stretch and crumple the icy crusts. Theoretically, there is not enough heat from the decay of radioactive isotopes in any of these satellites to do much work of this degree. A number of assumptions have been built into these models, however. The greatest uncertainty involves the non-ice, or rocky, components. It has been assumed by many that this rocky material is similar to the more volatile-rich meteorites that we find here on Earth, the carbonaceous chondrites. These are assumed to come from the outer parts of the asteroid belt and therefore to represent something of the composition of materials beyond Mars. These meteorites tend to have less of the radioactive elements that produce heat within the Earth than our own planet does, suggesting that the icy moons would get less heat and be cooler over time. One or more of these assumptions may be wrong. However, the diversity of geologic activity, with neighboring satellites being alternately ancient or volcanically resurfaced, suggests that something unique to each body has been controlling its evolution.

The first great Voyager discovery was active volcanism on Io (Fig. 5.3). At about the same time, Stan Peale and co-workers predicted volcanism on Io based on the tidal deformation and melting of this body as it orbits Jupiter. Tidal deformation has also been important on Europa, which has geologic evidence of a liquid water layer, and may have been important at one time on Ganymede. Io represents the most extreme example of tidal heating. The Earth–Moon tides are certainly important for controlling the flow of Earth's oceans, and may have helped promote the development of life along the oceans' shore. But aside from very small moonquakes, the tides have little direct influence on heat budgets within the Earth or the Moon. Not so with Io. The difference at Jupiter is that Jupiter itself is 1,300 times the mass of Earth, and, there are three additional large satellites in orbit near Io. The orbits of two of these, Europa and Ganymede, are currently timed to coincide with that of Io, such that each exerts a gravitational tug on the others. This distorts the orbits of all three, forcing them to be more eccentric or out-of-round than they would normally be. This orbital dance is called the Laplace Resonance.

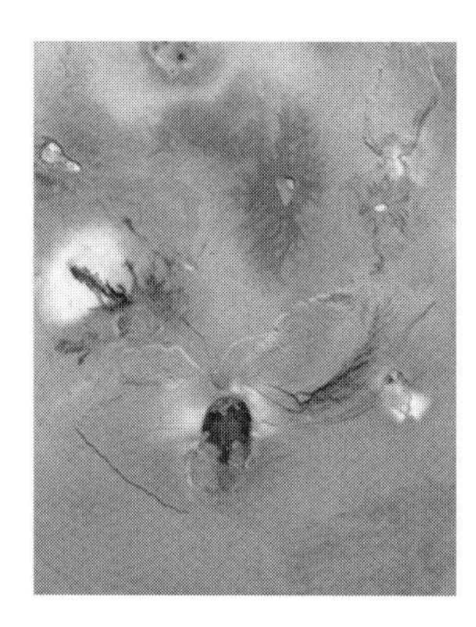

Figure 5.3 The volcanic surface of Jupiter's Moon-sized satellite Io. Gravitational tides raised by Jupiter and the other large satellites have partially melted the interior, leading to surface volcanism. The large mountain near center is 250 kilometers (405 miles) long.
Credit: Paul Schenk/LPI.

The forced eccentricity of Io's orbit changes its distance to Jupiter and results in large distortions in its shape over the course of a few days. The crustal tide on Io may be as high as 100 meters. By comparison, typical ocean tides on Earth are only a few meters (and locally tens of meters where funneled by submarine topography). This change generates tremendous amounts of frictional heat along microfractures throughout Io's crust and interior. Much of this energy goes into raising the temperature and locally melting the interior.

The effects of the tidal heating process may be directly related to the geologic history of all the galilean satellites. Io, the most strongly heated, is a volcanic maelstrom, covered with sulfur-rich volcanic deposits (Fig. 5.3). Some of these volcanoes may be erupting at 1,700 C, hotter than most lavas on Earth. Europa's tides are no more than 30 meters, but it is covered by a water layer that may be partially molten, and it has a young icy surface that may be currently undergoing deformation. Ganymede has an older surface and does not appear to be active today. However, it has been extensively resurfaced as recently as

1–2 billion years ago. This may have resulted from a temporary period of tidal heating associated with formation of an orbital resonance. Heating is no longer important on Ganymede, and was probably never important on Callisto, which shows no signs of geologic activity.

The icy satellites of Saturn and Uranus, especially Enceladus and Miranda which have total volumes roughly equivalent to the Antarctic ice sheet, are too small and do not experience nearly the same degree of tidal deformation that Io does. There does not seem to be the available tidal energy to melt large quantities of water ice on these bodies. The situation is eased considerably if we call on lower-melting point ices such as ammonium-hydrate, which has a melting point of only −97 C, rather than 0 C for water ice. Much less energy would be needed to melt some ammonium-hydrate, and tidal deformation may have been sufficient to do the job.

Circumstantial evidence for low-melting-point ices such as ammonia comes from the morphology of some of the volcanic deposits. Volcanoes on the icy satellites do not resemble the towering cones of Olympus Mons on Mars, or Mt Fuji on Earth. Extensive relatively smooth plains which resemble the lunar maria are observed on satellites such as Ganymede, Enceladus, Dione, and Triton. The most unusual features, however, are long symmetrical ridges on Miranda and Ariel, and small rounded domes, probably volcanic in origin, on Enceladus. The ridges can be a few hundred kilometers long and up to several kilometers high. The manner in which they lie on top of older terrains indicates that they formed by the extrusion of lavas along linear cracks in the crusts of Miranda and Ariel.

The morphology of volcanic deposits depends on the composition of the lava. Water is a very runny liquid compared with most molten rock. It generally cannot form thick or tall volcanic deposits or constructs such as we see on Earth or Mars. The volcanic deposits on Enceladus, Ariel, and Miranda form ridges 500–2,500 meters high. This is considered to be too thick to have been formed by a pure water liquid lava extruding on to the surface, which would flow easily and form smooth flat deposits. Something is needed to thicken the lavas so they form ridges and domes. Ammonia-water fluids, particularly if they are partially crystallized, as commonly occurs in terrestrial lavas, can be relatively thick and pasty. The presence of additional minor phases in the ammonia melt (perhaps methanol) also thickens liquid ammonia-water.

Whether or not these deposits were formed by liquid ammonia-water is unclear because we have not yet identified ammonia on these bodies telescopically. UV rays can break down ammonia into darker materials, thereby masking its presence spectroscopically. Whatever the composition of these deposits, they are more complex than pure water and indicate that the internal compositions of these satellites include a variety of other ices, including ammonia. These volcanic deposits are the only indirect evidence we have into the internal composition of the smaller saturnian and uranian satellites.

Ice worlds -- Oceanus Amokium?

Wisdom suggests that life requires liquid water to survive, or at least to develop. Recent magnetometer data acquired by Galileo suggests that liquid water oceans may be present within all three of Jupiter's large icy satellites. For Callisto and Ganymede, these water oceans may lie beneath several hundred kilometers of icy crust, but both geological evidence and magnetospheric interactions suggest that a liquid water layer up to 100 kilometers thick may lie just a few kilometers beneath the icy surface of Europa. Other large icy satellites, especially Titan, may have "wet" interiors as well. And, where there's water . . .

Europa – coming out of its shell?

The four large galilean satellites of Jupiter have been of seminal importance since their discovery in 1610. They represent a coherent family of objects whose basic properties tell a compelling story (Fig. 5.4). The density of each object decreases as we go outward from Jupiter, as the total water content increases. Contrary to some pre-Voyager expectations, geological complexity of these moons increases toward Jupiter. Io, the most geologically active, is waterless and composed mostly of rock and sulfur. The outer pair, Callisto and Ganymede, are relatively quiet geologically and are more than half water ice. Europa represents a transitional body in fundamental ways. Its density is three times that of water, indicating that it has some ice but not as much as Ganymede or Callisto. Europa has sufficient ice and/or water to form a layer roughly 150 kilometers thick surrounding a rock mantle and iron core. With its

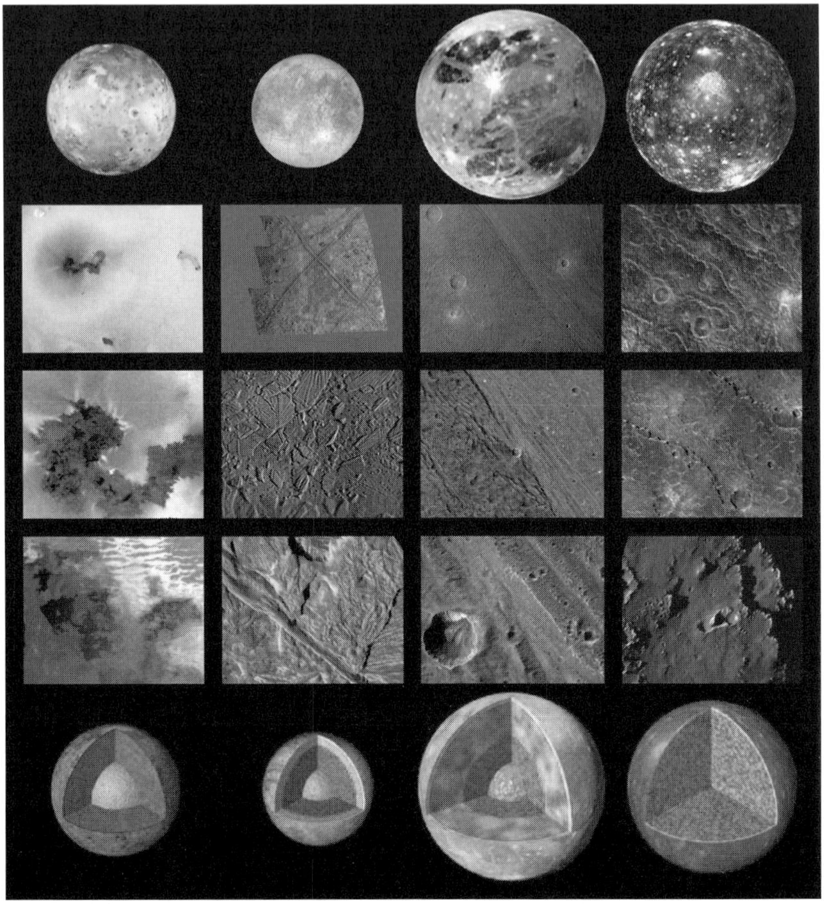

Figure 5.4 The galilean satellites dissected, from left to right: Io, Europa, Ganymede, and Callisto. Top four rows show increasingly higher resolution views of each satellite, from 10 kilometers to 1 kilometer, to 100 meters to 10 meters. Bottom row of cut-away diagrams shows our current understanding of the interior structure of each satellite. Io, Europa, and Ganymede probably have dense metallic cores. Europa, Ganymede, and Callisto have thin outer rinds, mostly of water ice 100 kilometers thick or so. Ganymede also has a water ice-rich mantle extending 800–1,000 kilometers into the interior and surrounding a rock and iron core.
Credit: Paul Schenk/LPI.

large rocky core and ice shell, Europa shares properties of both Io and Ganymede (Fig. 5.4, although its water content is between the two).

Voyager showed us that the surface of Europa is heavily fractured and sparsely cratered by impacts. Europa's icy crust is therefore geologically young and may be undergoing deformation today. The prediction and concurrent discovery of active volcanism on Io has made us realize that orbital tides can have a profound effect on the history of a planetary body. Indeed there has been speculation that Earth's tides have had a stimulating effect on the evolution of early sea life along the ocean shore. It was realized very early after the Voyager missions that the same tidal deformation melting Io should also provide an additional heat source for Europa's interior. Early estimates suggested that this tidal heating, though less than on Io, might be enough to melt much of Europa's icy outer shell, forming a global liquid water ocean beneath a frozen outer crust. A lot depends on exactly where Europa's water is and where most of the tidal heating takes place. The heating estimates remain very uncertain and may or may not suggest liquid water, but this has not stopped rampant speculation about subcrustal oceans and Europan fish.

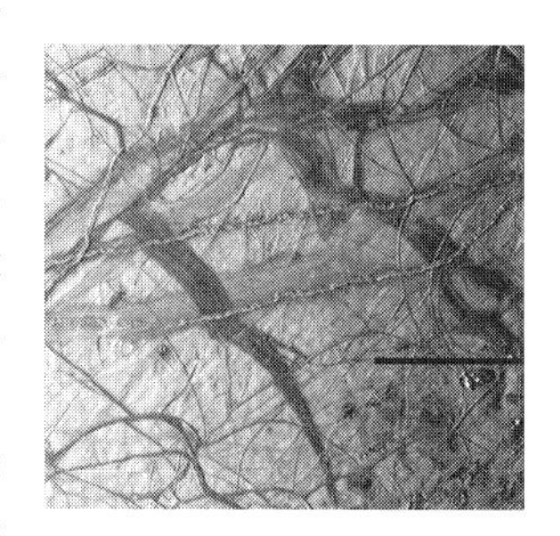

Figure 5.5 The intensely fractured surface of Europa. The dark wedge-shaped bands are 30–35 kilometers across. These bands formed when the icy crust fractured all the way through and split open. Dark icy material, either warm ice or refrozen liquid water from below then filled the cracks. *Credit:* NASA/JPL.

Although the surface of Europa as revealed by Voyager had a vague resemblance to floating sea ice, such comparisons are of limited value on an alien world. Two Voyager-based discoveries provided the first concrete evidence that an ocean might actually exist. The first discovery, by Paul Schenk, was that intact blocks of Europa's crust hundreds of kilometers across had been sliding past each other (Fig. 5.5). The second discovery, by Alfred McEwen, showed that many of Europa's largest fractures could be explained if Europa's icy shell was rotating nonsynchronously. (Like Earth's Moon, the jovian satellites keep one side always facing Jupiter, but, in the case of Europa, the crust appears to be very slowly drifting with respect to its core.) Both of these observations require that Europa's icy outer shell be decoupled or mechanically sliding on the interior. This can be achieved with either a layer of soft squishy ice or with liquid water. The question of whether an ocean existed on Europa awaited a closer examination of the surface.

The Great Leap Forward was provided by Galileo, which captured high-resolution views of the surface beginning in 1996. Galileo observed a very young surface (less than 100 million years old) and an array of geologic phenomena, all of which point to (but do not necessarily confirm) that a liquid water ocean exists beneath Europa's surface. Geologic evidence for a Europan ocean comes from a variety of unusual features that are unique to Europa. The most provocative of these features are the areas of "chaos." Chaos involves the breakup and disruption of older ridged plains (Figs. 5.4 and 5.6). Chaos is a highly disrupted and fragmented material and can include broken blocks of older icy crust. Also, many of Europa's tectonic features, including curved ridges, have been linked to tidal fracturing processes that seem possible only in a floating ice shell.

Another piece of evidence is the highly flattened appearance and unusual shapes of some of Europa's larger craters (Fig. 5.7). Flattening could be due to impact into a very thin crust overlying water. There are

Figure 5.6 Chaos on Europa. The crust on the left is mostly intact, except for a stray oval-shaped extrusion or uplift. But the right side is broken into a set of plates 5–10 kilometers across. These plates are separated by disrupted material known as matrix. Did these chaotic regions form when the icy crust melted or as the crust deformed convectively, due to high internal heat? Recent topographic analysis by the author suggests that the latter hypothesis is preferred. *Credit:* Paul Schenk/LPI.

also areas that appear to have been flooded by volcanic fluids (water in this case), but these areas are very small. Large 10 kilometer-wide oval features (Fig. 5.6) are evidence that the crust is convecting, somewhat like porridge will do if left on the stove too long. Unfortunately, all these geologic features might have been formed either in a very thin shell over a liquid subcrustal ocean or by slow flow within a layer of soft warm ice at the base of Europa's ice crust. To understand what is going on beneath Europa's ice crust requires corroborative evidence.

Galileo gravity measurements confirmed that most of the water is concentrated in the outer 150 or so kilometers, but were not detailed enough to tell us whether the water is frozen or liquid. Galileo's magnetometer has provided what may be the smoking gun – direct measurements of a possible ocean under Europa's icy shell. The magnetometer detected changes in the magnetic field that are consistent with the presence of an electrical conductor inside Europa. The logical choice for a conductor seems to be an ocean, and a potentially briny ocean at that. The dark material found associated with some of Europa's geologic features may ultimately be related to this briny ocean layer. Impacts, diapiric upwellings, or volcanic extrusions could all have brought oceanic material to the surface.

If Europa does have an internal water ocean (a Europa Orbiter may be sent ultimately to look for it), then the right combination of circumstances may be present to see the building blocks of life form. Heating due to tidal flexing during Europa's 3.6 day-long orbit must be responsible for keeping such a liquid ocean from freezing. This same tidal flexing may also be heating the rocky core and mantle of Europa. If so, the other ingredients of life, energy and food, could easily be added to the oceanic stew by way of volcanism on the ocean floor. The rocky

Figure 5.7a Tyre – a large impact crater on Europa. Although only 40 kilometers across, this is the largest known crater on Europa. Unlike normal craters on icy satellites (see Figs. 5.8 and 5.9) this impact, which would have been 40 kilometers across, instead formed a concentric set of fractures 140 kilometers across. Craters like these are evidence that Europa's shell is floating on a liquid water ocean, possibly 20 kilometers or more deep. *Credit:* Paul Schenk/LPI.

mantle is likely to contain some carbon compounds, which could be dissolved in the ocean as water percolates through the upper mantle, or as a direct result of volcanism.

The floor of Europa's ocean could potentially resemble parts of the Earth's seafloor along the great volcanic mid-ocean ridges. These ridges have given rise to colonies of non-photosynthetic organisms, including complex organisms, such as worms and crabs that rely on chemical energy from volcanic vents. What is not known is whether these types of life forms originated on the seafloor or migrated in from Earth's surface. Europa's crust is too thick now to allow light to penetrate to the ocean. If primitive life developed near the surface of Europa, it could easily have migrated to the floor of Europa's ocean and may still be there.

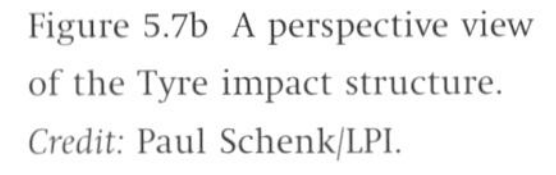

Figure 5.7b A perspective view of the Tyre impact structure. *Credit:* Paul Schenk/LPI.

So far, most of these ruminations remain in the world of speculation; some would say science fiction. A Jupiter moons orbiter should settle the question of whether the ocean exists today or did so in the past, and how thick the icy crust has been over time. Once that has been determined, then we must evaluate what the next step should be. Liquid water is only one element in the puzzle. Are there sufficient organic components within the ocean to make life or its organic precursors possible? Is there enough energy to sustain this chemistry against breakdown? Hopefully an orbiter can address some of these questions too, but it will be necessary to send a probe on to or into the crust to find out. Doing so without introducing contaminating terrestrial microbes, which we have discovered can live under rather extreme conditions, will be difficult!

Ganymede and Callisto – twins separated

The Voyager view of Ganymede, largest of the icy moons and next moon out from Jupiter, was of a satellite with a complex geologic record, but also of a world probably dead for a billion or more years. Galileo changed this impression radically (Fig. 5.8a). Ganymede has been

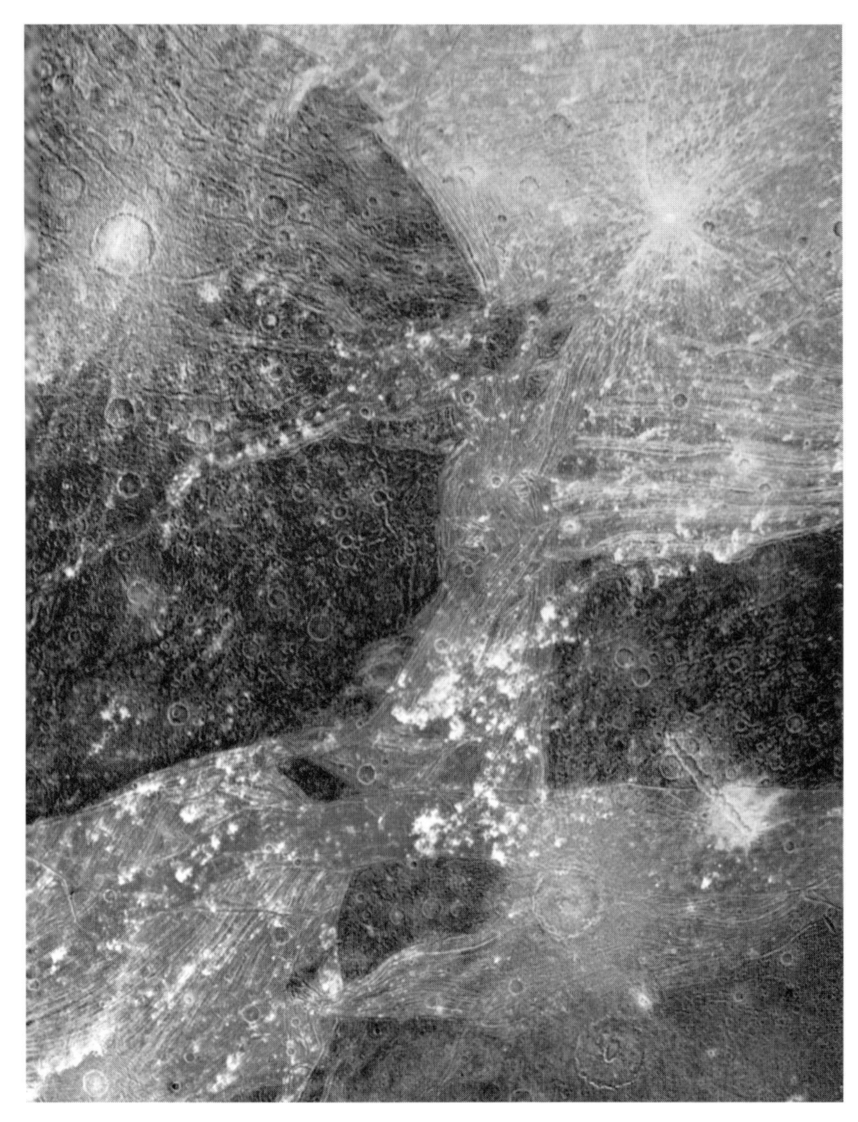

Figure 5.8a Regional Galileo view of Ganymede. Older dark terrains and younger bright terrains are apparent, as are numerous circular impact craters. Some bright terrains are heavily fractured. The linear chain of craters (a catena), at lower right, was formed by the impact of a comet that was disrupted when it passed very close to Jupiter. The bright floored impact crater at upper left is 85 kilometers across. The original images have a resolution of 1 kilometer per pixel. *Credit:* Paul Schenk/LPI.

revealed to be a planet in the truest sense. Gravity data from Galileo indicates that Ganymede is strongly differentiated, with an iron inner core, a rocky outer core, and an ice crust and mantle ~800 kilometers thick. Ganymede is the only satellite currently known to possess its own intrinsic magnetic field, roughly one tenth as strong at the surface as Earth's magnetic field. The location and formation of Ganymede's polar frost caps may be controlled by this magnetic field. A strong magnetic

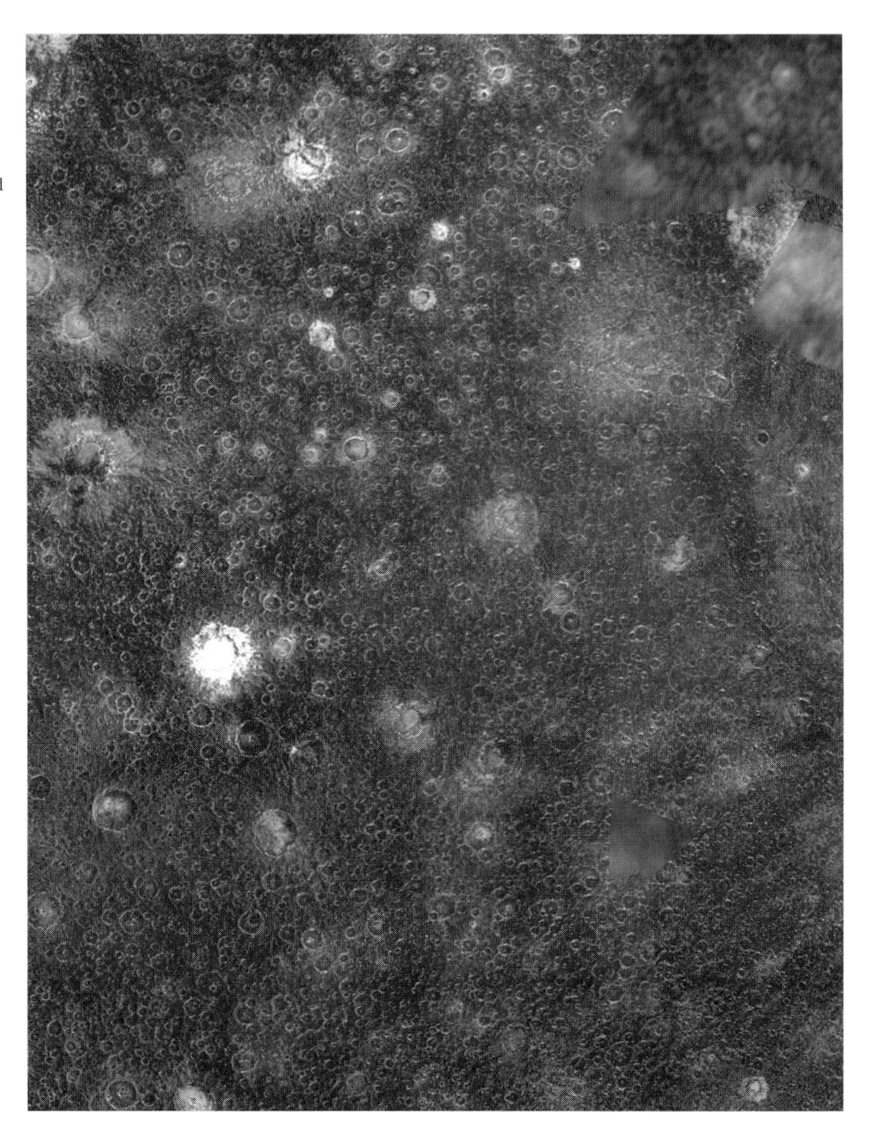

Figure 5.8b Regional Galileo view of Callisto. At this scale the surface of Callisto is dominated almost exclusively by innumerable impact craters. Image has similar resolution and scale as the Ganymede view in Figure 5.8a.
Credit: Paul Schenk/LPI.

field in a planetary body is usually evidence for convection or stirring in the interior. Subtle signatures in the magnetometer data also indicate that there may be a liquid water layer perhaps a few hundred kilometers below the surface. Although surface activity seems to have ceased some time ago, it appears that the interior of Ganymede is alive and kicking.

Geologically, 66% of Ganymede has been resurfaced by bright terrain (Fig. 5.8a). The rest of Ganymede is made up of older, more heavily

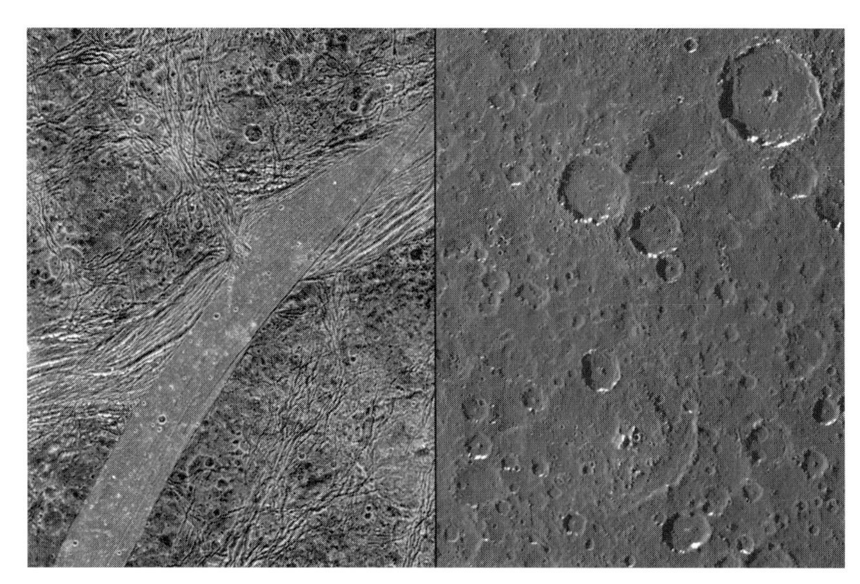

Figure 5.9 Close-up views of Ganymede (left) and Callisto (right). The fractured nature of Ganymede is apparent. Even older, dark terrains are locally fractured. The smooth swath of bright terrain across the image is approximately 25 kilometers across and was probably formed by extrusion of water lavas. Original images have resolutions of approximately 150 meters. *Credit:* Paul Schenk/LPI.

cratered dark terrain. The nature and age of the resurfacing is important to understanding the internal evolution of Ganymede. Many but not all bright terrains have been so heavily faulted as to erase all pre-existing structures (Figs. 5.4, 5.8a, 5.9). Despite this, most of the resurfacing now appears to have been accomplished, at least initially, by volcanism. This occurred when shallow structural troughs were flooded by low-viscosity lavas (Fig. 5.9). The brightness and strong spectroscopic signatures point to water or water-ice mixtures as the lava in question.

Callisto, on the other hand, is sometimes regarded as one of the least interesting, most boring, satellites in the solar system. Despite being similar in size, density, and overall composition to Ganymede, there is no preserved record of endogenic surface activity on Callisto. The dominant geologic activity has been impact cratering (Figs. 5.4, 5.8b, 5.9). Unlike Ganymede, Callisto is mostly undifferentiated, with no core or mantle. The brightness and morphology of young impact craters suggest the outer-most layers of Callisto are in fact ice-rich, but much of the rock and ice in the interior appears to be well mixed. This is consistent with the lack of an intrinsic magnetic field at Callisto. Despite this, there are indications from Galileo's magnetometer that parts of Callisto's icy interior, perhaps 200 kilometers deep, may also be liquid. No one knows what might be happening in these internal "oceans," or for sure if they exist, but it gives the imagination food for thought.

Extensive surface volcanism on Ganymede may be directly related to its internal and orbital evolution. Ganymede has always stood out in contrast to its twin, Callisto, and the Galileo gravity results make this contrast even more startling. If Ganymede was warm enough to separate into an icy mantle and iron core, then why didn't Callisto? Once again, the specter of tides rises. According to one theory, the three inner jovian satellites have passed through a series of tidal configurations. In one of them, Ganymede temporarily experienced unusual perturbations in its orbit. The resulting tidal heating may have been enough to melt part of Ganymede's interior. It is even possible that this melting triggered core formation (if it had not been formed already). Core formation in itself releases considerable amounts of heat and large parts of Ganymede's interior may have been melted. Callisto apparently has never experienced tidal deformation; the lack of geologic resurfacing may be evidence.

Although triggering of volcanism and tectonism by tidal deformation in the past forms an attractive coherent picture of how Ganymede evolved, there are loopholes. We do not know the age of the resurfacing with precision except to say it is probably at least 1.5 billion years old. Nor can we estimate with any great confidence when the era of tidal heating occurred on Ganymede, or the magnitude of any heating of the interior. Differentiation and melting of the interior may have occurred as a result of heating by accretional bombardment or radionuclide decay instead. Ganymede is 35% larger by mass than Callisto, and perhaps this was enough to cause a difference.

Figure 5.10 Crescent Titan. Sunlight refracting through the dense nitrogen atmosphere and organic haze creates the ring effect in this Voyager view from 1981.
Credit: NASA/JPL.

Titan – rain of goo . . .

Saturn's large moon Titan, the second largest satellite overall, may have some of the same ingredients of life as did early Earth, but in a different mix. Titan's methane-rich atmosphere was discovered by Gerard Kuiper in 1944, and Voyager discovered that nitrogen was the primary constituent (Fig. 5.10). In fact, Earth and Titan have the only large atmospheres in the solar system dominated by nitrogen. Some believe the early Earth may have been briefly shrouded in a methane haze similar

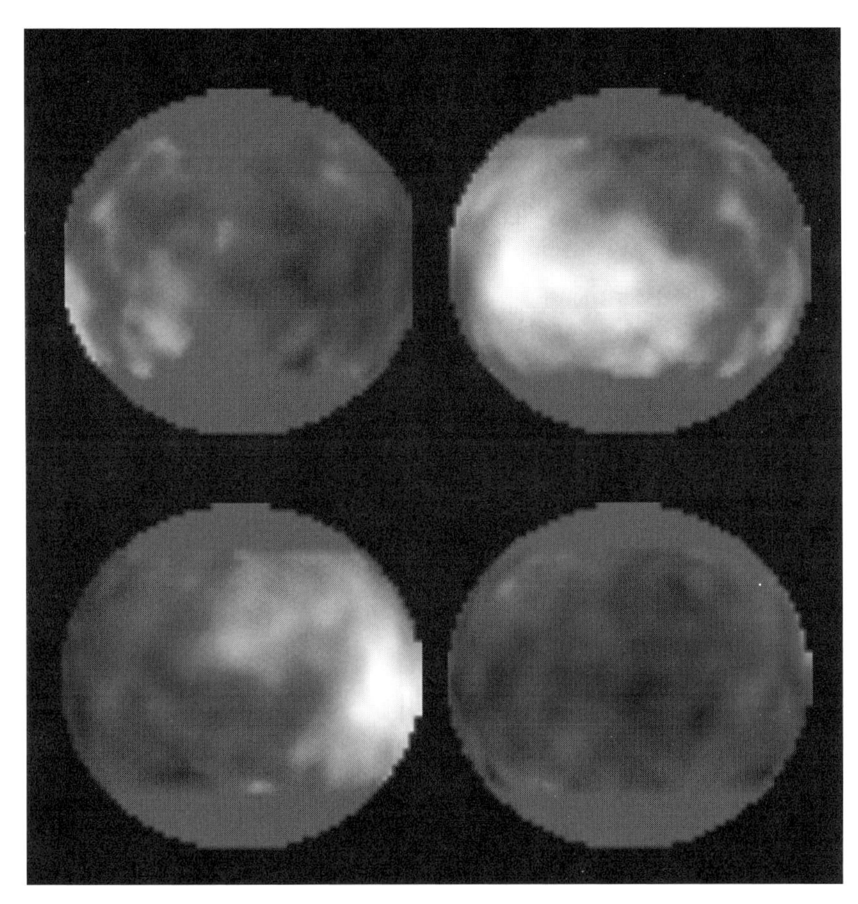

Figure 5.11 Hubble Space Telescope view of the surface of Titan. Although opaque to human eyes, the atmosphere is transparent in infrared wavelengths, something Cassini will use to advantage during its tour of the Saturn system. Although these views are not very high in resolution, they do tell us that the surface is not covered in a global ocean of organic goo. The bright spots may be volcanoes, mountains, or impact craters.
Credit: STScI.

to Titan's. Certainly, our atmosphere was also depleted of oxygen until 2 billion years ago, and methane was probably more abundant very early. Thus, a better understanding of the chemistry of methane-rich nitrogen atmospheres could be important for understanding our own planet's evolution.

Titan's methane should have been converted by sunlight to ethane and other organic molecules a long time ago. Heavier than gas, these molecules are supposedly raining out of the sky and forming a coating, perhaps even lakes on the surface of Titan. Dark areas in low-resolution global images of Titan (Fig. 5.11) are being interpreted by some, perhaps prematurely, as these lakes. If true, then the continued presence of methane in the atmosphere requires that it be resupplied from the surface in some manner. One mechanism is volcanism and the release of volcanic gases from the interior. Whether volcanism can in fact do the job is unclear, as is the validity of the entire ethane rain model. Despite this, one can easily envision a world with icy volcanoes, river valleys, and shallow seas. Voyager was unable to penetrate the haze, but the atmosphere is transparent at infrared wavelengths and, beginning in 2004, Cassini should reveal a very interesting planetary body.

Geologically, Titan is an almost blank page to us (Fig. 5.11). Geophysically, Titan closely resembles its large cousins Ganymede and Callisto, and is midway between them in size. If those two bodies have deep water oceans, as now appears likely, perhaps Titan does as well. Titan lacks its own magnetic field, however, and comparisons to Ganymede or Callisto are likely to be misleading. The presence of a methane-rich atmosphere certainly suggests that Titan has significant amounts of other volatile ices, such as ammonia, within its interior, perhaps more so than the galilean satellites.

Is Titan an early Earth in deep freeze awaiting a warming from a giant distended Sun in some far distant future? There are important differences. Titan is much colder, with a surface temperature of 94 K (−179 C). The crust is dominantly ices, not silicate or carbonate rocks as on Earth. We will have to await the new mission to Saturn to find the answers, beginning with an understanding of the basic geologic record and the internal construction of Titan. The global maps obtained by the Hubble Space Telescope and ground-based telescopes are too coarse to map the geology of the surface, except to say that it is diverse and consists of bright and dark areas. Radar and infrared mapping of the surface by Cassini should begin in 2004.

Triton

Neptune's solitary large moon Triton may be another example of tidal heating but with a twist. Triton orbits retrograde (backward) in a lopsided orbit (Fig. 5.12). Theoretical work by Bill McKinnon has shown that Triton was probably captured by Neptune from a solar orbit. In this sense it is probably a lost kin to Pluto and the other recently discovered Kuiper Belt objects (Fig. 5.12). Not that they would recognize Triton. The present tilted retrograde orbit of Triton is circular and is not influenced by tides. But the act of capture, which might have involved collision with a now-destroyed early moon of Neptune, followed by the circularization of the orbit appears to have devastated the wayward Triton. Calculations indicate that at least 90% of the interior was melted. This can only have resulted in the complete destruction of the original surface of Triton.

Voyager saw less than half the surface during its speedy flyby in 1989, but what it saw was beyond our wildest expectations. I remember watching as the images grew in detail with each passing day. It was only a week before the actual flyby that we finally knew how large Triton actually was. Voyager ultimately saw a very young surface, marred by very few impact craters. Recent estimates suggest that the surface may be less than 50 million years old, which would make Triton one of the youngest and most active bodies in the solar system.

Triton could easily be called the most geologically complex of the icy satellites. Extensive lava flows cover the surface (Fig. 5.13), as well as a variety of highly unusual terrain types. One of the most enigmatic is cantaloupe terrain (Fig. 5.14), noted for its complex pattern of ridges and closed depressions. This terrain probably formed when the crust overturned, forming a vast field of diapirs. Diapirs are roundish low-density blobs of solid crust that rise up from below to reach the surface (think lava-lamp). Voyager provided a surprise at Triton when it observed several geysers or plumes of dark material rising 8 kilometers into the thin nitrogen and methane atmosphere. Although it was once thought that these plumes might be triggered by solar heating of the surface, it now appears at least plausible that they could be generated by internal heat in some way.

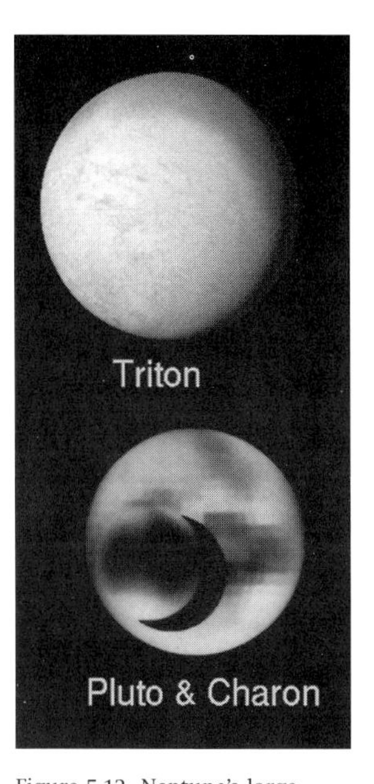

Figure 5.12 Neptune's large moon Triton (above) and Pluto (below) with its moon Charon. Although very similar in size and density and with very similar surface compositions, these two estranged twins have led very different lives. While Pluto and its moon have quietly orbited the Sun for several billion years since the collision that gave birth to Charon (see Chapter 6), Triton wandered too close to Neptune and was captured, with devastating consequences. Triton was observed by Voyager in 1989. Pluto will be visited for the first time during the middle of the next decade.
Credit: Paul Schenk/LPI.

The diversity of geologic features, especially volcanism, the estimates of surface age, and apparent inevitability of melting in the interior suggest that Triton is a geologically active world. It is therefore not only plausible but likely that Triton may be partially molten at depth and thus should be added to the list of icy bodies with subsurface oceans.

The density of Triton suggests that ices, presumably mostly water ice, make up approximately 30% of the interior. We also have direct evidence for exotic ice species on the surface. In addition to water ice, frozen carbon dioxide and the highly volatile frosts of nitrogen, methane, and carbon monoxide have all been identified (see the chapter by J. Stansberry in this volume for further discussion). With a frigid surface temperature of 35 K (−240 C), these gases have condensed as

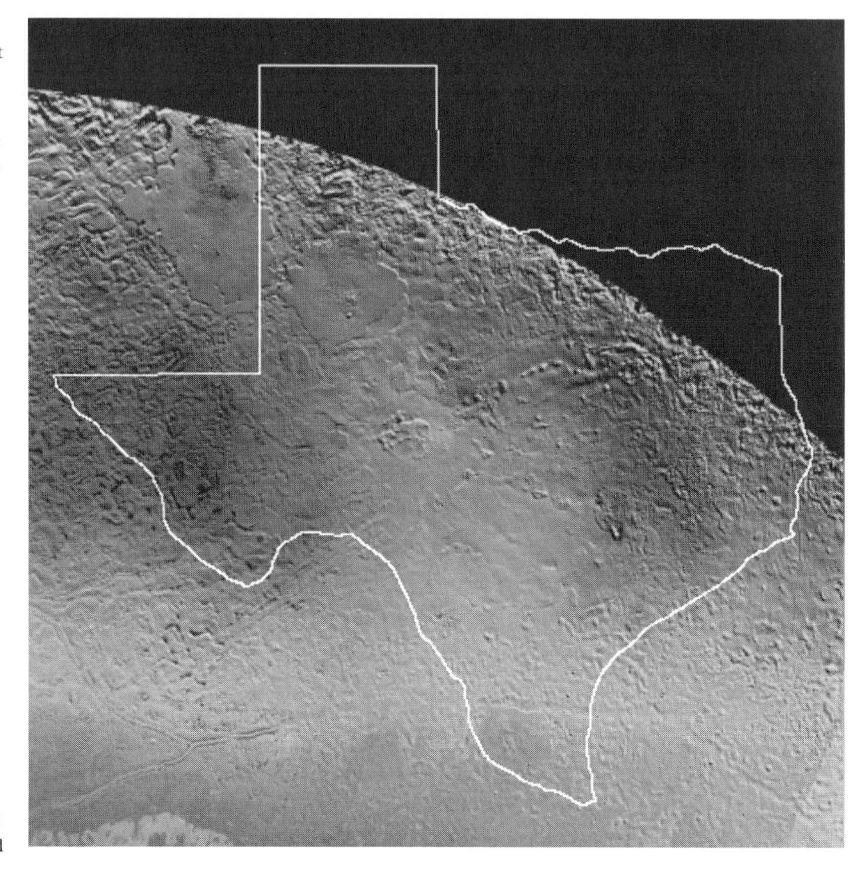

Figure 5.13 The volcanic plains of Triton. The volcanic lavas that formed these plains, pits, and calderas are believed to have been liquid ammonia water. The large volcanic crater near center is approximately 80 kilometers across. Outline of Texas is for scale.
Credit: Paul Schenk/LPI.

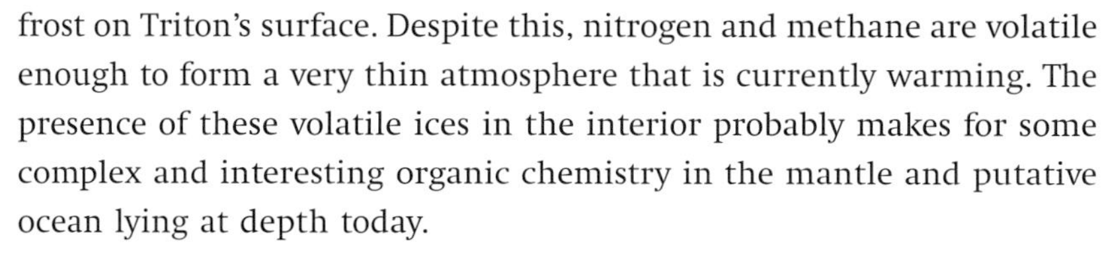

frost on Triton's surface. Despite this, nitrogen and methane are volatile enough to form a very thin atmosphere that is currently warming. The presence of these volatile ices in the interior probably makes for some complex and interesting organic chemistry in the mantle and putative ocean lying at depth today.

Figure 5.14 Cantaloupe terrain on Triton, formed by massive diapiric (convective) overturn of Triton's icy crust. Each cell is on average 40 kilometers across and could swallow a major city. Mosaic is a combination of medium- and high-resolution Voyager images, with best resolution of 400 meters per pixel.
Credit: Paul Schenk/LPI.

It's a not-so small world after all

The large satellites of Jupiter and Neptune are not the only satellites influenced by the tides. The small satellite Enceladus may have been partially melted due to its resonance with other saturnian satellites. The same may be true of the uranian satellites Ariel and Miranda. Even Dione, Tethys and Titania show limited evidence for resurfacing The importance of tidal distortion and heating in planetary bodies is one of the lasting legacies of the Voyager mission.

Figure 5.15 The divergent twins Mimas (left) and Enceladus (right). These small saturnian satellites are only 400 and 500 kilometers across, respectively. Mimas is mostly heavily cratered, with a few fractures. Enceladus is mostly smooth, with some irregularly shaped craters that may be volcanic in origin, or at least modified by volcanism. *Credit:* NASA/JPL.

Enceladus, Miranda, and Ariel

Most of Saturn's middle-sized icy satellites have been partially cratered, and appear to be relatively old. Enceladus, second out from Saturn and only 450 kilometers across, is relatively crater free (Fig. 5.15). Ridges and banded terrains similar to those found on Ganymede cross its surface. Neighboring Mimas is almost exactly the same size, yet it is a boring nearby cratered lump of ice (Fig. 5.15). Like Ganymede and Callisto, we are faced with a paradoxical set of twin satellites.

A clue may be in the location of Enceladus in the center of the distended and faint E-ring of Saturn. There are no obvious reasons why any particular moon of Saturn should have a ring associated with it. The ejection of dust due to meteorite bombardment would presumably affect other satellites as well. The geologic youth of a moon surrounded by a swarm of tiny particles may not necessarily be a coincidence. Some craters on Enceladus are partially filled with steep-sided domes uncharacteristic of impact events and more like volcanic extrusions (Fig. 5.15). It is therefore possible that volcanism is ongoing and is the source of the small particles that populate the E-ring. Here lies yet another challenge for the Cassini mission.

Even after our experience with Enceladus, the complexity of the tiny uranian moon Miranda, one of the smallest of the spherical icy satellites, came as a surprise (Fig. 5.16). Because of the lopsided tilt of the Uranus system (the pole of rotation is tilted more than 90 degrees in comparison with the other planets), Voyager was able to see only the

Figure 5.16 Uranian satellites Miranda (left) and Ariel (right). Like Enceladus, the complex geologic history of these small bodies was unexpected. Tidal heating and low-melting point ices, such as methane or ammonia, may be part of the explanation. Miranda is 470 kilometers across; Ariel 1,170 kilometers across.
Credit: NASA/JPL.

southern hemispheres of the uranian satellites in 1986. As a result our geologic maps and histories of these satellites are somewhat incomplete. Nonetheless it is readily apparent that Miranda is not a simple moon (Fig. 5.16). While half of it is cratered plains, the remainder is divided into at least three large tracts of resurfaced terrains called corona. Corona are banded and generally much smoother and slightly darker than the cratered plains. Stereo images clearly show high scarps along some of these bands however, indicative of faulting.

Ariel is even more amazing in some respects than Miranda. Most of it has undergone some form of geologic resurfacing (Fig. 5.16). We see evidence for large fault-bounded canyons and smooth plains, implying rifting of the crust and flooding of the surface by lavas. These canyons are up to 5 kilometers deep. There are even some blocks of Ariel's crust that have broken and tilted. Clearly Ariel has been partially melted and its crust started to break up.

What caused Miranda's split personality? One hypothesis suggested that Miranda was blown apart and then reassembled. Miranda is close to Uranus and relatively small. The large mass of Uranus tends to focus asteroids and comets toward it and increase their relative velocity. Because the outer uranian satellites are heavily cratered, we can infer that Miranda gets blasted even worse by objects potentially large enough to fragment it. If Miranda has a rocky core, then the darker corona features could be evidence of the reaccretion or reaccumulation of fragments of a proto-Miranda shattered during one of these events.

More recently, detailed analysis of the fault patterns and directions shows their formation is more consistent with material rising from the core of Miranda. Ice blobs rising from the inside of Miranda would result in stretching and faulting of the overlying crust and associated volcanism. It now appears that, if Miranda was shattered and reformed, we do not see direct evidence of it in the geologic record. Rather we see evidence that Miranda started to turn itself inside out.

What then caused both Ariel and Miranda to start turning inside out, and the possibly ongoing volcanism on Enceladus? Miranda is not currently in resonance with the other uranian satellites, as Io is with the other jovian satellites. It is therefore not experiencing tidal heating. A hint may be in that Miranda's orbit is slightly tilted with respect to the other uranian satellites. We now suspect that Miranda and Ariel were both locked in transient orbital resonances with Umbriel or one of the other outer uranian satellites in the past. The slightly tilted orbit of Miranda may be a remnant of that time. The process of squeezing and melting to form volcanic features is easier if you have ammonia-water in the interiors, which is softer and much easier to melt than plain water ice. Similarly, tidal interactions with Dione or another satellite may provide the heat to melt Enceladus. Although these models are attractive, theoretical difficulties remain in explaining how this heat is distributed and indeed whether it is enough to do the job.

Iapetus – come over to the dark side

As we have seen, the surface of Iapetus is highly unusual, divided between very dark and very bright icy hemispheres (Fig. 5.2). Spectra of the dark side may match carbon and organic material. The dark hemisphere of Iapetus is also the one that faces forward as it rotates around Saturn. Some have speculated that the darkening comes from sweeping up of small dark particles blasted loose from the outer saturnian satellite Phoebe. An alternative is that organic material has been blasted or eroded off the top of Titan's atmosphere and has struck Iapetus. This scenario is similar to our understanding of how Io's sulfur-rich volcanic material has apparently been trapped in Jupiter's radiation belts, only to pelt Europa's surface later.

These explanations appear to work in principle but are not entirely satisfactory. No other Saturn satellite shows such a strong brightness pattern, even though Rhea and Dione are both physically much closer to Titan. Also, the boundary between bright and dark is not gradual but is sharply defined and easily mapped on the images. This favors an explanation involving resurfacing, perhaps by some volcanic material. Why it should be dark is unclear, except to note that ammonia and other volatile ices darken when exposed to sunlight or radiation for long periods. If volcanic, why then is dark material limited to one hemisphere? Basaltic lavas flooded only one side of our Moon; perhaps something similar happened on Iapetus. Perhaps bombardment of dust from Phoebe does play a role, wearing Iapetus down and forming a broad depression which lavas later filled. Perhaps a large impact formed a deep crater, which later became the source of lava. Of course, the dark material may be ancient and the current orientation of Iapetus may have no relation to how the dark material originally formed. The answers must await the arrival of Cassini in 2004.

We have learned much about the crazy worlds of the outer solar system over the past 25 years. Yet a fundamental understanding eludes us. No samples have been returned from any of these bodies, and our estimates of surface ages are uncertain by at least 50% in most cases. As the Mar Exploration Rovers to Mars in 2004 demonstrate, we will have to go to surfaces of these bodies directly to determine their composition, age, and geologic history. We need not go to every satellite, just a few representative bodies. The Galilean satellites are key, as is Titan. Plans are being made to return to Jupiter sometime later this century. The Huygens probe due to land on Titan in early 2005 will be a start, although it was not designed to survive a landing on a hard surface. Hopefully the descent imaging and atmospheric measurements will whet our appetites for more.

JOHN A. STANSBERRY
Steward Observatory, University of Arizona

6

Triton, Pluto, and beyond

The edge of the known solar system lies roughly 40 astronomical units from the Sun (1 AU = 150 million kilometers; the distance between the Earth and the Sun), not too far beyond the orbit of Neptune, which orbits the Sun at 30 AU. Pluto is the best-known and the largest (known) object in this region, but it is just one of many bodies orbiting beyond Neptune, and one of the nearest to the Sun. Neptune's large moon, Triton, is akin to Pluto in size, composition, and even in its origin, although one is a planet and the other a planetary satellite.

More than 700 other bodies also have orbits larger than that of Neptune, all of them discovered since 1991. These new members of our stellar system, referred to as Kuiper Belt, Kuiper Disk, Edgeworth–Kuiper Belt, or trans-Neptunian objects, are being discovered at a rate that doubles the number of known objects orbiting beyond Neptune each year. While considerably smaller than the planets, the trans-Neptunian objects are typically quite large (typical diameter >100 kilometers) compared to their cousins the asteroids (typical diameter 50 kilometers) in the inner solar system. Pluto, Triton, and the trans-Neptunian objects are unlike the rocky bodies orbiting between the Sun and Jupiter, and the gas-giant planets Jupiter, Saturn, Uranus, and Neptune. These unexplored worlds at the fringe of the solar system are composed of roughly equal parts water ice and rocky material; they have very thin or nonexistent atmospheres.

All of these bodies formed from the mixture of gas and dust that coalesced into the flattened, rotating cloud of material known as the proto-solar nebula. Most of the mass of the disk was quickly concentrated into the center and became the Sun. A small fraction of the nebula was spun out into a thin disk that contained a higher proportion of solid matter relative to gas than the original cloud, the proto-planetary disk.

The bulk composition of all the bodies in the solar system is approximately consistent with the high temperatures in the inner regions of the disk, and the cold temperatures in the outer regions, which would be the case as the early Sun heated the inner regions much more effectively than the outer regions. Only materials with high melting and vaporization temperatures (so-called refractory materials that typically form rocks) condensed in the inner disk, while in the outer disk substances with low melting and vaporization temperatures (so-called volatile materials, most importantly water) were able to condense.

Thus the inner planets and the asteroids are composed primarily of rocky materials, while at Jupiter and beyond objects are much richer in water (primarily as ice). Because water and its constituent atoms hydrogen and oxygen are very abundant in nature relative to refractory substances, the bodies with a lot of water are also more massive than their rockier cousins. The gas giant planets grew so fast and to such large masses by accumulating water that they were then able to gravitationally trap huge amounts of hydrogen gas from their vicinity of the proto-planetary nebula – gas that had managed to escape being incorporated into the Sun. In fact, the gas giants were large enough to form their own small spinning disks of material – planetary sub-nebulae – during formation.

Somewhere not far beyond the position where Neptune formed out of the proto-planetary disk, objects still accreted considerable amounts of water during formation, but they stopped attracting the huge amounts of gas that now dominate the atmospheres of the planets Jupiter through Neptune. Pluto, the next planet beyond Neptune, has only 1/10,000 the mass of Neptune, and has an atmosphere so thin (about 1/10,000 the pressure of Earth's atmosphere) that it was not detected until 1989. By contrast, about one twelfth of Neptune's mass resides in its predominantly hydrogen and helium atmosphere, which has a pressure about 200,000 times that of Earth's atmosphere.

Pluto's brethren, Triton, and the trans-Neptunian objects, are also small, water-ice-rich bodies. The cause of this transition from massive gas giants to small icy bodies is not fully understood. Most likely it is due to a combination of the fact that planets formed more slowly in the outer solar system (due to slower orbital dynamics) and the fact that the gas in the proto-planetary disk was eventually cleared out as the early solar wind developed. Yet it was this apparent edge to the solar system that led Kenneth Edgeworth (in 1949) and subsequently Gerard Kuiper (in 1951) to hypothesize the existence of many undiscovered objects orbiting the Sun beyond Neptune. Thus the names Edgeworth and/or Kuiper are sometimes used in referring to the trans-Neptunian objects being discovered today.

Pluto's story

Pluto is singular in some respects among the nine known planets – some even decline to classify it as a planet, instead categorizing it along with the asteroids, comets, and trans-Neptunian objects, as a minor planet. It is the smallest of the planets. With a diameter between 2,300 and 2,400 kilometers it is one half the size of Mercury. Pluto's orbit is about six times more elliptical than the orbits of the other planets (save Mercury), and for a period of about 17 years during each orbit of the Sun, Pluto lies closer to the Sun than Neptune (Pluto passed back outside the orbit of Neptune in June of 1998). Pluto's orbit is also inclined by 17 degrees to the plane of the solar system, as compared with typical inclinations for the other planets of about 2 degrees. (In fact, Pluto's orbit is the archetype for the orbits of a class of trans-Neptunian objects.) On the other hand, Pluto's diameter is about two times greater than that of the largest asteroid (Ceres, diameter = 912 kilometers) and the largest known trans-Neptunian object (Quaoar, with an estimated diameter of 1,100 kilometers), and is correspondingly eight to 15 times more massive than those very largest examples of the minor planets. Pluto's moon, Charon (pronounced sharon), at 1,200 kilometers diameter, is also larger than the largest known minor planets, and is the largest moon in the solar system relative to its primary. Unlike all the known minor planets, Pluto also possesses a bound atmosphere, at least for the 50 years or so of its orbit when it is relatively close to the Sun (as it is now).

Whether Pluto is a planet is neither very important nor interesting from a scientific perspective. However, the story of how Pluto came to follow its orbit, possess its large moon, and share many characteristics with Triton and the trans-Neptunian objects reveals much about the history of the outer solar system, and sets the stage for discussing the icy nature of all of these bodies (Fig. 6.1).

According to current theory regarding the formation and early evolution of the outer planets, Pluto did not form in the orbit where we now find it. Neither, for that matter, did Uranus or Neptune, nor did Triton form in orbit around Neptune, nor did many of the trans-Neptunian objects form in the orbits they currently follow. Very early in the history of the solar system, the proto-Uranus and proto-Neptune arose from the random coalescence (accretion) of small bodies (planetesimals) orbiting the early Sun in the proto-planetary disk. Their growth did not

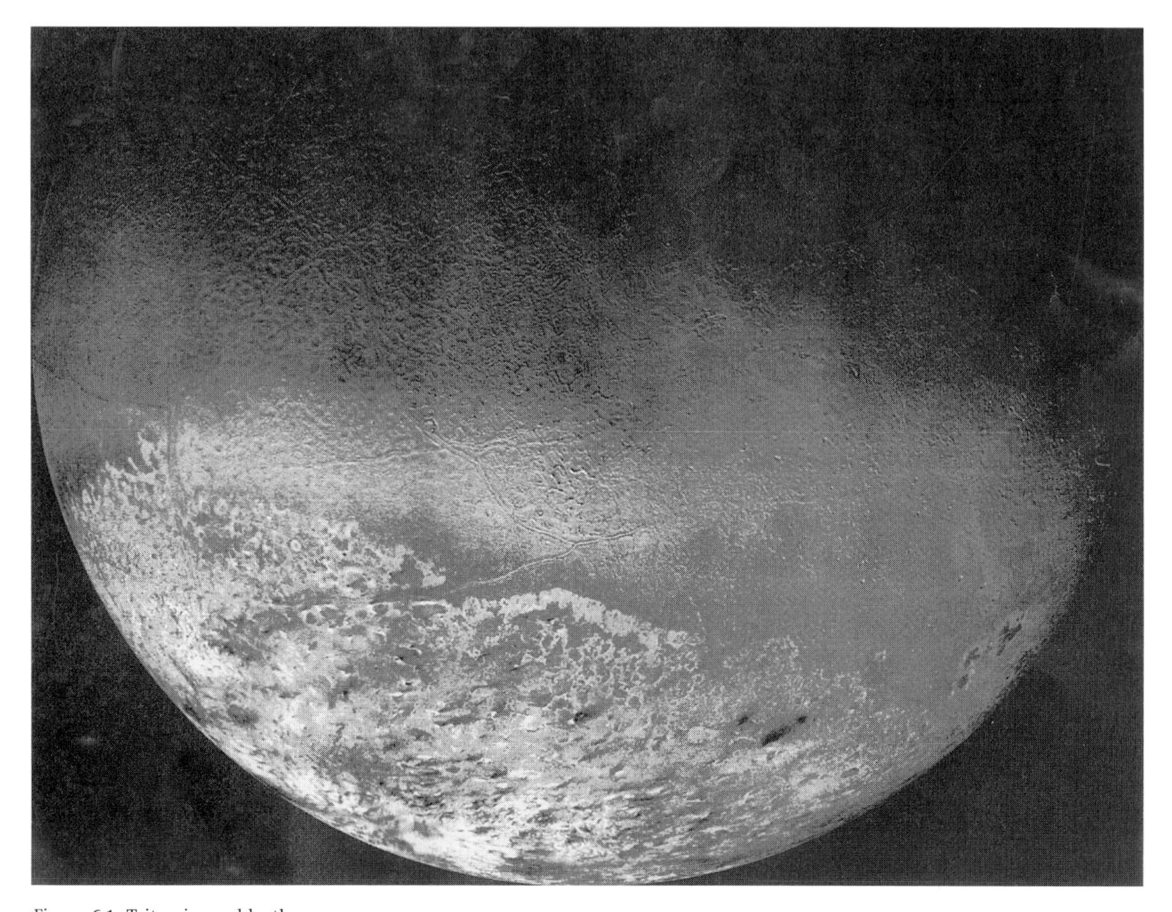

Figure 6.1 Triton imaged by the Voyager 2 spacecraft in 1989. The bright mottled and streaked deposit at the bottom of the image is the South Polar Cap (SPC). The chevron shaped deposit just below and to the right of the image center lies just south of Triton's equator. The bright wispy material above (north) of the SPC may be freshly deposited nitrogen ice. Cantaloupe terrain dominates the upper left area. Ridge systems cross the entire surface, while in the upper right water lava appears to have inundated much of the surface. *Credit:* NASA and USGS, Flagstaff, AZ.

immediately exhaust the solid material orbiting near them within the disk. As they continued to grow by capturing and consuming other objects, many of the objects within their spheres of gravitational influence escaped accretion. For many of these objects the joy of not being consumed by their neighbor was short lived: they were promptly sent on to new orbits crossing the orbit of the proto-Jupiter, and were either consumed by Jupiter, or had their orbits further perturbed such that they were plunged into the Sun or sent on to huge, highly elliptical, randomly inclined orbits. These scattered asteroids and planetesimals compose what is known as the Oort Cloud, which is the source of long-period comets (such as Hale-Bopp).

Uranus and Neptune were not unscathed by this process: because they preferentially tossed the smaller objects from their regions of the disk inward towards Jupiter, they consequently moved outward in order to conserve orbital angular momentum. While each of the encounters between Uranus or Neptune and a smaller body only changed the orbit of the planets by a tiny amount, the cumulative effect of many, many such encounters over a period of several million years was to cause these huge planets to migrate outward through the solar system. Neptune is estimated to have gravitationally scattered roughly ten Earth-masses-worth of leftover planetesimals, and to have migrated outward about 7 AU from the position where it first formed. The migration took about 50 million years to complete – a comparatively short time when compared with the age of the solar system (4.6 billion years).

As Neptune migrated outward through the proto-planetary disk it encountered and gravitationally interacted with many planetesimals that would otherwise have continued in their orbits unperturbed. Most of these bodies suffered the fates described above, but a small fraction of them, including Pluto, became trapped in an orbital "resonance" with Neptune. Resonances occur when two solar system bodies move in such a way that the ratio of the periods of their individual motions is equal to a ratio of two small integers. For example, the orbital period of Pluto is 248 years, Neptune's is 165 years, and the ratio of Pluto's to Neptune's period is within 1% of the ratio 3:2. Even though the orbits of Pluto and Neptune cross, the two never collide because of their resonant orbital periods.

Two bodies that are in resonant motion will tend to remain in resonant motion, even if an outside force perturbs one or both of them. Detailed calculations and computer simulations show that once Pluto became trapped in the 3:2 resonance with Neptune, it would be carried outward along with Neptune as it migrated, and, over time, Pluto's initially circular, zero-inclination orbit would be transformed into an elliptical, inclined orbit such as it follows today.

Thus Pluto, like Neptune and the other planets, formed in a nearly circular orbit essentially coplanar with the orbits of the other planets. Pluto's current orbital characteristics are a consequence of a general rearrangement of the orbits of Uranus, Neptune, and essentially all of the objects that had orbits between the positions where Uranus and Neptune began to form and about 40 AU. In fact, the resonant, inclined, elliptical orbits of Pluto and some of the trans-Neptunian objects are the smoking gun indicating that Neptune underwent considerable orbital migration. There is very little chance of even one object ending up in the 3:2 resonance by means other than described above, and the presence of many (potentially thousands) of objects in that and other resonances with Neptune essentially require orbital migration as the explanation.

Triton and the Trans-Neptunian objects (TNOs)

The presence of Pluto in the 3:2 resonance with Neptune by no means precludes other objects from becoming trapped in the same resonance during Neptune's orbital migration, nor is the 3:2 resonance the only resonance capable of trapping objects. Of the 700 or so currently known trans-Neptunian objects (TNOs), roughly one in four has an orbit resonant with Neptune. These objects make up the "resonant" class of TNOs. Most of the resonant TNOs are in the 3:2 resonance, like Pluto, but a comparable number are in the 5:3 resonance (Neptune orbits five times each time they orbit three times) with orbital radii close to 42 AU. And a few objects fall in the 2:1 resonance, with orbital radii near 48 AU. The resonant objects have elliptical, inclined orbits like Pluto's.

Pluto's moon, Charon, is a special member of the TNOs in the 3:2 resonance, because it almost certainly did not form in orbit around, and contemporaneously with, Pluto. It is much more likely that, like the Earth's Moon, Charon formed as the result of a collision between the planet and another comparably sized body. Such a collision forms a disk of ejecta debris in orbit around the planet, and that debris subsequently re-accretes to form a moon.

The other major class of TNOs are the "classical" TNOs. These objects have nearly circular, low inclination orbits, with orbital radii in the range 42–48 AU (although the upper end of this range is probably due to the difficulty of discovering objects orbiting farther out than 48 AU, and not to a real lack of objects at larger radii). The classical TNOs are thought to have essentially the same orbits today as when they formed over 4 billion years ago. The third class of TNOs are the "scattered" TNOs. This class comprises a handful of objects with large eccentricities and inclinations, and orbital radii much larger than the other TNOs (note that an object with a large orbital radius can still come into the 30–40 AU region of the solar system if its orbit is also highly elliptical).

The scattered TNOs may be objects that have been recently perturbed by a gravitational interaction with Neptune.

Triton is perhaps the ultimate resonant object, lying as it does squarely in the 1:1 resonance with (in orbit around) Neptune. This points towards Triton's real birth as an object that formed within the proto-planetary disk, on a circular orbit about the Sun, and in all likelihood very similar to Pluto. Unlike Pluto, the early Triton was not captured by a resonance external to Neptune's orbit, nor was it scattered inward or outward by a gravitational interaction. Instead, Neptune simply ran over Triton during its orbital migration, capturing it into orbit.

The primary evidence supporting this hypothesis is Triton's retrograde (opposite to Neptune's spin), inclined orbit. While many small moons of the gas giant planets have retrograde orbits (and are thought to have been captured), Triton is the only large moon with a retrograde orbit. Another scenario proposed some 20 years ago proposed that Pluto and Triton changed places: Pluto started as a moon of Neptune, and Triton as a small planet. Triton had a close encounter with Neptune, where it had a very close encounter with Pluto, ejecting Pluto from orbit about Neptune and capturing Triton into its retrograde orbit. Modern computer simulations show that the probability of this scenario happening is essentially zero. Furthermore, Pluto's orbit, and the orbits of many of the TNOs, have now found a natural explanation in the orbital migration of Neptune (Fig. 6.2).

Like the formation of Charon by a giant impact on Pluto, the capture of Triton into retrograde orbit around Neptune was probably an extremely violent affair. Neptune almost certainly formed with a retinue of moons, just as Jupiter, Saturn and Uranus did. Each of those planets possesses a system of more than ten moons, some of which are more than 1,000 kilometers in diameter, in prograde orbits (orbiting in the direction of the planet's spin) that lie close to the equatorial plane of the planet, with orbital radii of about 3–30 planetary radii. These systems of "regular" moons formed at the same time as the planet they orbit, and formed from the planetary sub-nebula of gas and dust. Today, Neptune has a system of only six regular moons, the largest of which is only 400 kilometers in diameter, and all have orbital radii less than five Neptune radii.

These unusual characteristics of Neptune's regular satellites are consistent with disruption of an early, more typical system of moons by a

Figure 6.2 The positions of the known KBOs (circles) and former KBOs that have been perturbed on to orbits interior to Neptune's (Centaur objects, triangles). Open symbols indicate recent discoveries. Circles with crosses and the large gray ellipses show the positions and orbits of the giant planets. Pluto is the circle with the cross near bottom center, apparently on the orbit of Neptune, which itself appears at the lower right.
Credit: Figure courtesy of the Minor Planet Center of the International Astronomical Union.

catastrophic event of some kind. The capture of Triton out of a solar orbit into a retrograde orbit about Neptune would have been disruptive indeed. For starters, the leading theory for how the capture might have been initiated is that Triton made a close pass through the Neptune system and collided with a pre-existing moon with a diameter of 10–400 kilometers. Such a collision would have slowed Triton down enough to keep it from escaping Neptune's gravity field. This collision would not have completely destroyed Triton, but would have deposited a huge amount of heat in the outer layers – enough heat to melt all the ice in Triton, producing a 300 kilometers deep, globe-enveloping ocean of water (with a thin ice crust). The capture would have left Triton in an initial orbit around Neptune that was highly elliptical. As Triton orbited it would have continued to pass close (approaching to within 5–10 R_N) to

Neptune, perturbing the orbits of the remaining regular moons. These would have fallen into Neptune, collided with each other, or they may have collided with Triton. In any case, the capture wreaked havoc both on Neptune's regular satellites, and on Triton itself. As described below, Triton's orbit eventually would have evolved into its present-day circular form, with yet more catastrophic consequences.

Pluto, Triton, and the trans-Neptunian objects all formed in circular orbits around the Sun. They formed from the same proto-planetary disk of material as the other planets, but probably formed a bit more slowly, remaining too small to accrete much hydrogen and helium from the disk before those gases were blown out of the solar system. The present-day circumstances of Pluto and Triton and at least the resonant class of TNOs were profoundly influenced by the early orbital migration of Neptune.

Triton and Pluto: twin siblings of a distant Sun

Triton and Pluto are very similar in many ways (see the Table 6.1 summarizing some of their characteristics), including their bulk, surface, and atmospheric compositions, surface temperatures, and sizes. One of the characteristics that clearly differentiates them from the inner planets is that they are composed of about one-third water ice by mass (or nearly one half ice by volume). Compositionally, this makes Triton and Pluto more similar to the icy moons of the giant planets than to the terrestrial planets. Because of their large size and violent histories, both Triton and Pluto are probably "differentiated," that is the materials in their interiors have been segregated into layers according to density. Thus both have an outer layer (about 300 kilometers thick) composed primarily of water ice, with some other volatile molecules (for example, carbon monoxide, carbon dioxide, methane, and nitrogen) mixed in with and/or forming a thin veneer of ice at the surface.

Geology recorded in water ice "rock"

At the temperatures of Triton and Pluto water ice is as hard as rock is on Earth, and, because it is the major material in their outer layers, the geology of these objects is recorded in ice, just as the geology of Earth, Mars, Venus, Mercury, and the Moon is recorded in stone. Very little is

Table 6.1

Characteristic	Triton	Pluto
orbital period, solar (year)	164.8 (Neptune)	247.7
orbital period, satellite (days)	5.88	6.39 (Charon)
rotation period (days)	5.88	6.39
distance from Sun (min:max) (AU)	29.76 : 30.37	29.65 : 49.42
moon–planet distance (R_{planet})	14.3	16.7
orbit ellipticity (circular = 0)	<0.0001	0.005
radius (km)	1353	1175 ± 20
density (g/cc)	2.06	≊1.9
rock : ice : iron mass fractions	65% : 30% : ≤5%	70% : 30% : 0%
surface composition	N_2:H_2O:CH_4:CO_2:CO	N_2:H_2O:CH_4:CO
surface albedo (low:high)	0.2 : 0.85	0.1 : 0.85
surface temperature (low: high) (K/°C)	38:55 /−235:−218	38:60 /−235:−213
Atmospheric pressure (bars, Earth = 1bar)	0.000016	≅0.00002
Atmospheric composition	N_2:CH_4:CO	N_2:CH_4:CO

known about the geology of either Triton or Pluto – what is known is derived from images of Triton taken by the Voyager 2 spacecraft when it flew through the Neptune system in late 1989.

One of the most fascinating discoveries made by Voyager was that Triton's surface appears to be very, very young by solar system standards. As with all of planetary geology except that of the Moon (for which we have collected samples) determining just how young Triton's surface is in absolute terms (years) is currently impossible. In general this is because we lack physical samples of the surfaces of the planets. In the particular case of Triton, and other icy bodies, even collecting samples of the surface may not help, because there is no known chemical signature that can be uniquely linked to the age of water ice (in rocks isotopic chemistry is used to compute ages). However, it is clear that

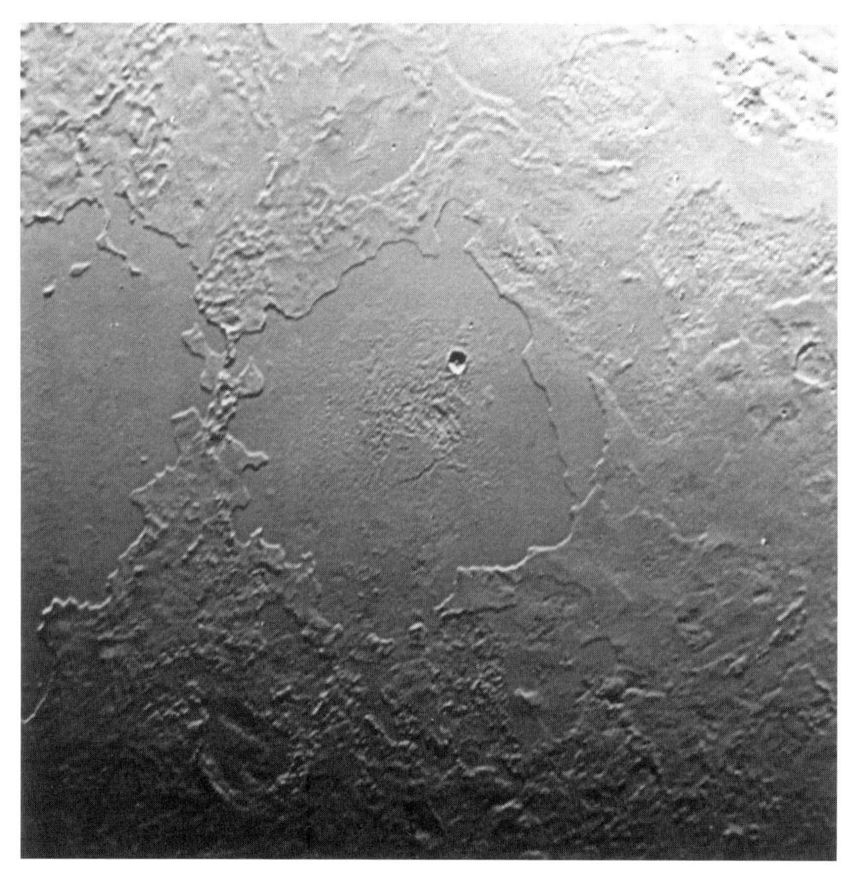

Figure 6.3 This Voyager image shows a volcanic caldera on Triton. The caldera is flooded with cooled "lava" of water. A few impact craters are visible as round depressions. The lumpy region near the center of the caldera is presumably the source region for the water lava. *Credit:* NASA and USGS, Flagstaff, AZ.

the surface of Triton is very young relative to other icy bodies because of the nearly complete lack of impact craters on Triton (Fig. 6.3).

Impact craters are a ubiquitous feature of most solid-surfaced bodies in the solar system. Impacts were an essential part of the formation process for all planets, moons, and minor planets. The collisions of the final stages of formation, as well as a continuous sprinkling of impacts that continues today, have left their imprint in the form of densely packed craters on the surfaces of all of them. The lack of craters on a surface indicates that later processes buried, eroded, or otherwise erased them. On Earth, erosion, plate tectonics, sedimentation, and volcanism have all played important roles in largely erasing the cratering history of our planet. On the Moon, volcanism has flooded the tremendous mare basins (themselves impact craters), burying the craters that littered their floors. The Martian highlands are heavily cratered, while the northern plains have relatively few craters due to the effects of erosion, deposition, and volcanism.

Smaller bodies in the solar system tend to have well-preserved crater populations because they lack the internal heat sources that drive volcanism and tectonism, and they lack atmospheres and the accompanying erosive effects. The smaller moons of Jupiter, Saturn, Uranus, and Neptune tend to display heavily cratered, ancient surfaces. The large moons of the jovian planets display a variety of crater densities, both when compared with each other and, in some cases, when different regions of a single object are compared. The youngest surface from this group of objects is that of Io, Jupiter's volcanically hyperactive moon, which does not have a single identified impact crater. Roughly tied for second place are the surfaces of Triton and Jupiter's moon Europa. Jupiter's Callisto is very heavily cratered, while its neighbor Ganymede is in parts heavily cratered, but in other regions almost crater free.

Tidal evolution and giant impacts

Why does Triton have such an apparently young surface? In the case of Io and Europa young surfaces are not a big surprise because they are locked in an orbital dance with Jupiter and each other, causing their orbits to be elliptical. Jupiter's gravity distorts the shape of Io and Europa (that is, it raises a "tide") by a significant amount, and that distortion changes rhythmically as they approach and recede from Jupiter as they move around their orbits. The time-varying nature of this tidal flexure leads to heating of their interiors, just as a piece of wire heats up if one bends it back and forth repeatedly. That heating expresses itself as profuse volcanism on rock-rich Io, and as a combination of faulting, deformation, and water volcanism on ice-rich Europa. Triton's surface exhibits broad areas that have been inundated by water "lava," faults, and regions that appear to have been strongly deformed. All of these features are nearly free of craters, indicating that they were formed in the not too distant past. This evidence indicates that some heating mechanism operated in Triton's interior fairly recently. Yet Triton is not currently subject to a tidal heating mechanism like Io and Europa because its orbit is very circular.

Triton would have experienced considerable heating after its capture into orbit about Neptune. As noted earlier, Triton would have been in an initial orbit that was highly elliptical, with closest and farthest orbital distances of about 7 and >500 Neptune radii respectively. This initial orbit was then transformed, over a period of only a few thousand years, into the current circular orbit with a radius of 14 Neptune radii. This orbital evolution occurs because of tidal distortions of Triton's shape caused by Neptune's gravity. The tidal distortion was large when Triton was close to Neptune, and was small when Triton was far from Neptune, and so varied as Triton followed its elliptical orbit. This tidal flexing led to heating in Triton's interior (just as Io's interior is being heated today), and caused the evolution of Triton's orbit from elliptical to circular. This tidal heating would have been augmented by heat generated as Triton collided with Neptune's system of moons, in a series of giant impacts.

One of the important differences between Triton and Pluto is their density. The density difference may not appear to be large, but it provides an insight into the different formation histories of these two bodies. On Triton, either tidal evolution or collisional heating alone would have provided enough energy to completely melt all of the ice. Pluto was not subjected to strong tidal heating (although Pluto's moon Charon was), but the giant impact that formed Charon would have deposited enough energy to melt all of the ice in Pluto. The additional heat available at Triton may have raised temperatures there high enough that a significant amount of its water ice not only melted, but also vaporized. The water vapor would, along with other substances like nitrogen, and carbon dioxide, have formed an atmosphere for a time after Triton's capture. Because Triton has a weak gravitational field, some of this early atmosphere would have escaped into space, leaving Triton with more dense, rocky material than Pluto has. An alternative explanation is that the different densities of Triton and Pluto are a result of their formation. Because they probably formed in almost the same region of the solar nebula and are nearly the same size we might expect them to have formed with very similar densities, so this scenario is perhaps less likely than the former one, but still cannot be ruled out.

Is Triton's apparently young surface explained by tidal heating and impacts with Neptune's pre-existing moons? The true explanation will not be known until we learn much more about Triton's geology and interior from an orbital science mission. Any explanation that results from such a mission will be incomplete unless it provides an explanation of the density contrast between Triton and Pluto, and should also take into account (currently unavailable) knowledge about the age of Pluto's surface.

Kuiper Belt objects: cousins to Triton and Pluto

As noted earlier, the Kuiper Belt objects (KBOs) formed in the same region of the solar system as Pluto and Triton. It is reasonable to expect that KBOs are composed mostly of water ice and rocky material, and water ice has indeed been identified on the surfaces of two KBOs. Less clear is whether KBOs should contain abundant methane, nitrogen, and carbon compounds. Many such materials are highly volatile – that is, they tend to vaporize even at the low temperatures found in the outer solar system. Once a volatile material vaporized, the gases would easily escape from KBOs due to their small size and weak gravitational fields. The composition of KBOs has undoubtedly been modified over the past 4.6 billion years, just as the compositions of Triton and Pluto have probably been modified.

Triton and Pluto today

Both Triton and Pluto had dynamic histories around the time of their formation and subsequent evolution on to their current orbits billions of years ago. Unlike the asteroids, most of the moons of the giant planets, and the KBOs, Triton and Pluto are still undergoing constant change even today.

The focus of change on these objects now is in their atmospheres and on their surfaces, which undergo strong seasonal cycles. In fact, the seasonal changes that are occurring involve both the surfaces and the atmospheres in such an interrelated way that Triton and Pluto can be thought of as having coupled surface–atmosphere systems.

Nitrogen, methane, and atmospheres

Studies in the early 1980s of the spectrum of sunlight reflected from Pluto and Triton in the early 1980s revealed strong absorption features due to methane. Initially it was not known if the methane was in the solid, gas, or liquid phase, but before too long it was widely accepted that the absorption features on both bodies were caused by methane ice on their surfaces. Based on rough estimates for the surface temperature, it was thought that methane formed an atmosphere on both objects.

The paradigm for the atmosphere of Triton was complicated considerably in 1986 when, also using spectroscopy, nitrogen was discovered. While the absorption feature due to nitrogen was quite weak, its very existence implied that the surface of Triton was strongly dominated by nitrogen, not methane as had been assumed. This follows from the fact that methane is a very strong absorber of light (small amounts of methane create strong absorption features), while nitrogen is a phenomenally weak absorber. In fact, the nitrogen absorption is a so-called forbidden band; nitrogen can only absorb light in that band if it is in a very dense (effectively solid or liquid) state. The discovery also meant that Triton's atmosphere was dominated by nitrogen because the vapor pressure of nitrogen is around 10,000 times greater than that of methane at a given temperature. In spite of the obvious importance of the discovery of nitrogen to understanding conditions on Triton, it proved frustratingly difficult to decide whether the nitrogen was present as a solid or as a liquid.

A primary factor contributing to the difficulty of determining the physical phase of Triton's methane and nitrogen was that the temperature of Triton's surface was unknown. Up until 1989 scientists believed that Triton was roughly twice as large as it actually is. That misconception was based on the assumption that the reflectivity (or albedo) of Triton's surface was about 0.2, which is typical for outer solar system moons. By combining this estimated albedo with the known distances from the Sun and Earth, Triton's radius was estimated to be 2,600 kilometers, and its surface temperature to be about 60 K (−213 C or −351 F). That is well below the 90 K melting point of methane, so it seemed clear that the methane would be present as an ice on the surface and as a gas in the atmosphere. The case was not as clear for nitrogen, which has a melting point of 63 K. The nitrogen absorption feature mentioned above seemed to require the nitrogen to be in a very transparent form, probably liquid, with nitrogen gas present but undetected in the atmosphere. Speculation about methane-covered "landmasses" and nitrogen oceans made Triton seem like it might be a very exotic place indeed (Fig. 6.4).

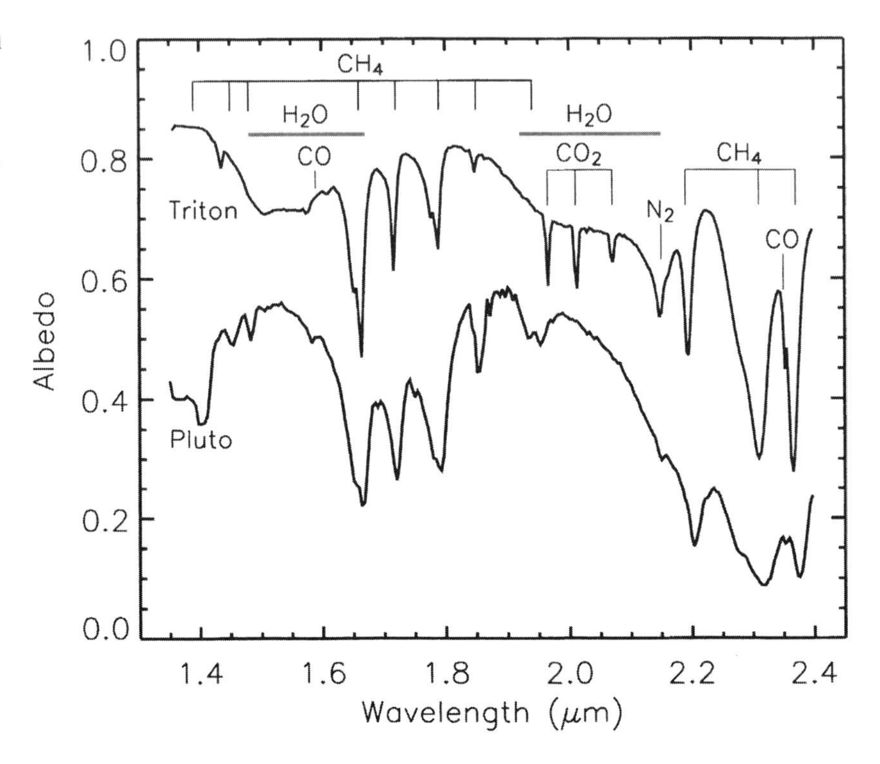

Figure 6.4 Spectra of Triton and Pluto at near-infrared wavelengths showing the complex absorption bands of methane, water, carbon dioxide, and carbon monoxide. *Credit:* Dale Cruikshank/NASA and William Grundy Lowell Observatory.

During the summer of 1989 it became apparent that Triton was more reflective (and therefore colder and smaller) than had been previously thought. The Voyager 2 spacecraft was in the final approach to its August encounter with the Neptune system. Based on the 2,600 kilometers predicted radius, images were taken several weeks prior to the encounter with the expectation that they would show Triton's resolved disk. However, Triton was still an enigmatic point of light in those first images, and remained point-like in images taken in subsequent weeks. When images from Voyager did finally show a resolved Triton, it was only half as big as predicted. This first direct measurement of Triton's size also implied that it was nearly four times as reflective as had been assumed, and that the surface temperature was a frigid 40 K – well below the melting point even of nitrogen. It was suddenly clear that the nitrogen and methane that had been spectroscopically detected a decade or so earlier were both present on the surface as ices.

At nearly the same time that the Voyager spacecraft flew through the Neptune system, gathering a wealth of scientific information, a fortuitous alignment of orbit and stars allowed Earth-bound scientists to collect a few minutes' worth of data that have revealed a great deal about Triton's twin, Pluto. The alignment was predicted well in advance, and indicated that Pluto would occult (that is, block the light from) a relatively bright star, and that during the occultation Pluto's star-shadow would sweep across the southern Pacific Ocean. Occultation predictions are notoriously tricky, but in the case of the 1989 occultation the predictions for the path of Pluto's shadow across the Earth turned out to be accurate.

A team of scientists used a telescope mounted in an airplane (the Kuiper Airborne Observatory, since decommissioned by NASA) to fly into the shadow-track at just the right time and they obtained very detailed measurements of the brightness of the occulted star as Pluto passed in front of it. Unlike the relatively sharp drop in brightness to be expected if Pluto was an airless body, the measured light curve showed a gradual decline indicative of an atmosphere. The rate of decline in the star's brightness as it was obscured by Pluto was a direct measurement of the slope of the density vs. altitude curve of the atmosphere. That slope is related to the atmospheric temperature by the molecular mass of the gas in the atmosphere. Assuming that Pluto's atmosphere was methane gave an atmospheric temperature between 55 and 60 K, consistent with the temperature of warm areas on Pluto's surface.

The occultation light curve was not a simple, smooth decline: part way through the gradual decline in the star's brightness, a sudden increase in the rate of decline was observed. A layer of haze or cloud could have caused this "kink" in the light curve, but so could a strong temperature gradient in the atmosphere. A definitive answer to what caused the kink may never be known – as will be discussed below, Pluto's atmosphere has strong seasons, and may be quite different by the time a spacecraft can visit or a better stellar occultation observation made. However, the discovery of a second icy constituent on Pluto provided a strong indication that the kink was caused mostly by temperature structure, not haze or clouds.

In 1992, nitrogen was discovered on Pluto. Unlike the uncertainty about whether Triton's nitrogen was liquid or gas, on Pluto it was immediately thought that the nitrogen was present as an ice. Three factors contributed to the quick identification. Better laboratory measurements of the absorption feature of nitrogen showed that the ice absorption feature was quite distinct from that of the liquid, and the ice feature was a better match to the spectrum of Pluto. Second, having studied Triton, it was fairly clear that nitrogen ice would be mostly in bright (high albedo) areas on the surface, and would therefore be colder than if it were in the warm areas. The temperature of high albedo areas on Pluto was estimated to be about 40 K (well below the melting point of nitrogen), so the nitrogen was almost certainly present as ice. But the key to deciding the nitrogen was in its icy form was the fact that in 1992 we already knew Pluto's diameter, albedo, and even the distribution of light and dark areas on its surface, unlike the pre-Voyager situation for Triton. (How we knew all that about Pluto will be discussed shortly.)

The presence of nitrogen ice on Pluto immediately suggested that the atmosphere would be almost entirely nitrogen, because of its large vapor pressure. This change in our understanding of the composition of the atmosphere required a revision of the atmospheric temperature derived from the occultation light curve – using the molecular mass of nitrogen instead of methane gave an atmospheric temperature of 100–105 K, much warmer than any area on Pluto could be. An atmosphere that warm could only connect to the surface by a temperature inversion – warm above the surface, and steadily colder towards the surface. This was exactly the kind of temperature gradient that would have produced the "kink" in the stellar occultation light curve, so there was no particular need to have a haze layer to make the kink. In 2002, Pluto occulted two more stars. The analysis of the data from these occultations is still very preliminary, and it is impossible to confidently say more than that the atmosphere appeared to be different than it was in 1989, consistent with the idea that strong seasonal variations probably take place on Pluto. However, it is not even clear if the surface pressure increased or decreased, nor is it clear what the implications of the new data are for Pluto's radius.

How were Pluto's diameter, regional surface markings, and albedo determined without sending a spacecraft? In 1978, Christy discovered Pluto's moon Charon while observing at the US Naval Observatory W, just a few miles from Lowell Observatory, where Pluto itself was discovered. Within a few years it was apparent that Charon's orbit was oriented such that Charon appeared to oscillate up and down, rather then going around Pluto: the orbit was nearly edge-on as viewed from Earth. As Pluto and Charon moved slowly around the Sun, Charon's

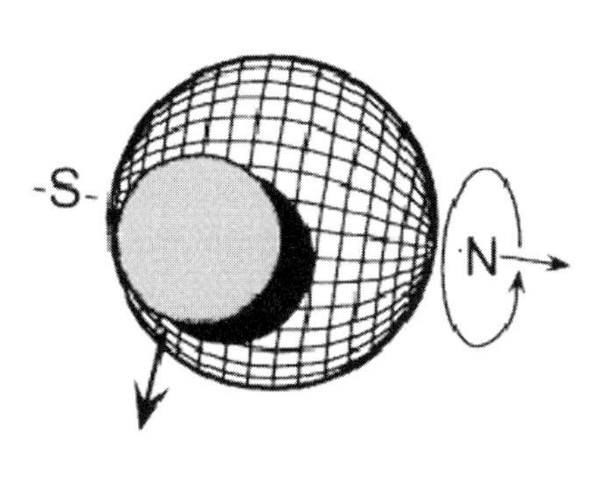

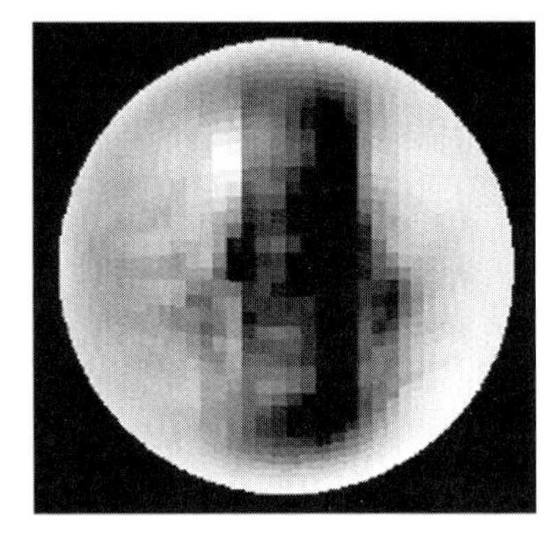

Figure 6.5 The geometry of a mutual event between Pluto and its moon Charon (left), and a map of Pluto created by mathematically modeling observations of many mutual events.
Credit: Eliot Young, Southwest Research Inst. Boulder, CO.

orbit became even more edge-on, and, in 1985, Charon began to alternately pass in front of and then behind Pluto once every 6.4 day orbit. When Charon passed in front of Pluto, it blocked out our view of part of Pluto's surface, and when it passed behind, Pluto blocked our view of part of Charon's surface. By carefully measuring the combined brightness of Pluto and Charon before, during, and after these eclipses (or mutual events) it was possible to determine the albedo of the part of their surfaces that had been eclipsed (Fig. 6.5). While each event lasted only five minutes or so, the events went on for six years because of the slow orbital motion of Pluto around the Sun. Maps of the surfaces of Pluto and Charon were created by combining data from observations taken over the entire mutual-event period. Pluto's and Charon's radii were also determined to good precision from these data sets, and once their sizes were known their albedos were also known, as described earlier for the case of Triton.

Ice transport and seasons

The nitrogen and methane ices on Pluto and Triton are mobile: they sublime (turn directly into gas) in areas with strong sunlight, and recondense from the atmosphere in areas with little or no sunlight. Just as evaporation of water from your skin cools you, sublimation of nitrogen or methane ice cools sunny areas on the surfaces of Pluto and Triton. Likewise, just as water vapor condensing on a cold glass warms it (and its contents), condensation of nitrogen and methane gas to form ice warms poorly illuminated and dark areas on Pluto and Triton. As the ice sublimes or the gas condenses "latent heat" goes along with the molecules moving from one phase to the other. The atmospheres of Triton and Pluto act as transport systems, absorbing the gas subliming from ice in the summer hemisphere, and supplying gas in the winter hemisphere where nitrogen and methane ice are condensing. Because of the latent heat associated with sublimation and condensation, the atmosphere also transports energy across the surface.

The familiar behavior of water and its phases has a further similarity to Pluto's and Triton's nitrogen and methane ice–atmosphere systems. If a pan of water and ice cubes are put on a stove, the temperature of the water remains at 0 C as long as there is any ice remaining in the pan. This is because heat coming from the burner is entirely used up in melting the ice, rather than heating the water already in the pot. Similarly, once water begins to boil, a thermometer placed in the pan reads 100 C regardless of how high the stove is turned up: all of the heat is used up transforming the liquid water to gas (steam).

On Triton and Pluto sunlight falling on nitrogen or methane ice does not cause it to warm up – instead the extra energy is completely used up changing the ice to gas. The same is true for the reverse processes both for water (e.g. freezing of ice-laden water occurs at 0 C until all of the liquid is gone) and for nitrogen and methane. On the night side and in the winter hemisphere of Pluto and Triton, ice forms on the surface, and the surface cannot cool off until ice stops condensing on it. The atmosphere is like a very long pot of ice water, with one end on the stove and the other end in the freezer. Heat flows in at one end, and ice melts; heat flows out the other end, and water freezes; and, in spite of all of the heat moving from one end of the pot to the other, the temperature of the water and ice in the pot is 0 C everywhere. The same is true for Triton's and Pluto's atmospheres: heat flows into the ice–atmosphere pot in the summer hemisphere, and ice sublimes. Heat flows out of the system in the winter hemisphere, and ice forms; and the temperature of the ice and atmosphere is independent of location to such an extent that the temperature of the un-illuminated poles of Triton and Pluto is equal to the temperature of the most intensely lit areas in their summer hemispheres.

The analogy between Pluto's and Triton's atmosphere–nitrogen ice system and the water–ice system is not perfect. The difference is that on Triton and Pluto the pressure of the atmosphere adjusts itself in response to the ice temperature. If most of the nitrogen ice is in the summer hemisphere the atmospheric pressure is larger than if most of the ice is in the winter hemisphere. The pressure of the atmosphere depends on the temperature of the nitrogen ice: the atmosphere is in vapor-pressure equilibrium with the ice on the surface. The ice temperature depends on the balance between the amount of solar energy falling on the nitrogen ice and the amount of thermal energy re-radiated by the ice. Because all of the ice is at a single temperature the energy terms have to be added up over all of the nitrogen ice everywhere on the surface in order to compute the ice temperature. These energy terms depend on location on the surface. Sunlight is strongest during daytime and in areas where the sunlight strikes the surface vertically, while there is no sunlight at night or above the winter arctic circle. The thermal re-radiation term depends on the temperature of the ice and a re-radiation efficiency factor called emissivity. Just as a surface with a low albedo is a good absorber of sunlight, a surface with a high emissivity is a good emitter of light. Given two surfaces with different emissivities but with the same albedo, the high emissivity surface will be colder than the low emissivity surface. Most surfaces are good emitters, with emissivities of 0.9–1. Nitrogen ice has an emissivity around 0.8 on Triton and Pluto at this time.

The atmospheric pressure on Triton and Pluto depends on the temperature of the nitrogen ice, which in turn depends on the ice albedo, emissivity, and on the distribution of the ice on the surface. As mentioned above, nitrogen ice is mobile, subliming from sunny areas and re-condensing in dark areas. The depth of ice that can sublime or condense in one Earth year is about one centimeter. Over a Pluto or Triton year (248 or 165 Earth years) roughly a meter of nitrogen ice can be deposited on, or removed from, an area. The ability of nitrogen ice to re-arrange itself means that the seasonal behavior of the ice-atmosphere systems of Pluto and Triton are potentially rather complex.

It is impossible to understand how atmospheric pressure will change over time without being able to predict the distribution of ice on the surface. In principle, making such predictions of the ice distribution should not be difficult, and beginning around the time of the Voyager encounter with Triton several computer programs were created to compute the distribution of nitrogen ice as a function of time by using

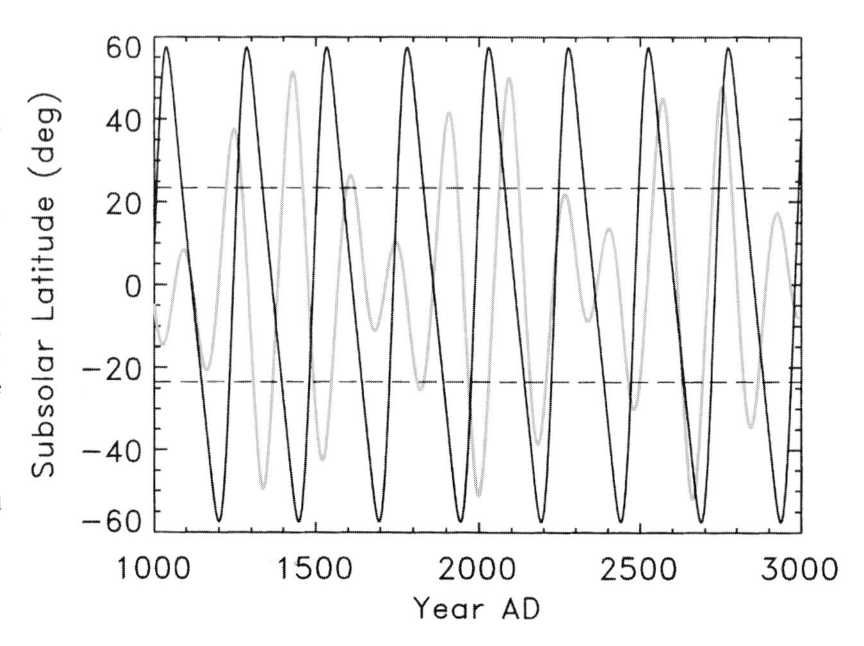

Figure 6.6 The latitude of the subsolar point on Triton (gray line), Pluto (black line), and Earth (range indicated by dashed lines) as a function of time. On Earth the subsolar point is restricted to 23 degrees latitude, with positive and negative maxima corresponding to northern and southern summer. On Triton the subsolar point can be as much as 50 degrees (north or south of the equator), and the variation is complicated. On Pluto the subsolar point moves in a regular pattern, varying over nearly 60 degrees north and south from the equator. The uneven spacing of the peaks in the Pluto curve is caused by the eccentricity of its orbit.
Credit: Pluto data courtesy NASA/JPL Horizons Service.

knowledge of the position of the Sun as a function of the season, and keeping track of nitrogen sublimation and condensation. The problem is that the computer models do not seem to explain what we believe we see on Triton: they tend to predict a shrinking southern polar cap, and a northern polar cap extending well down towards Triton's equator. What Voyager seems to have imaged was almost the reverse of what the models predicted.

Figure 6.6 shows the position of the Sun in the skies of Triton and Pluto vs. time. The Triton curve results from the combination of the inclination of Neptune's spin axis (30 degrees), and the tilt of Triton's orbit (and spin axis) to Neptune's spin axis (23 degrees). Triton's orbital plane precesses because of the tidal influence of Neptune (just as Earth's spin axis precesses under the tidal influence of the Sun), so that sometimes the orbital tilt and Neptune's tilt add up, while a precessional cycle later, the two tilts almost cancel. On Earth, the Sun moves between ±23 degrees north and south latitude each year. On Triton, the motion of the Sun can be as large as ±50 degrees during some Neptunian years, but during other years it is quite small. On Pluto, with its spin-axis tilt of 123 degrees, the Sun passes almost over the poles at summer solstice. Pluto has an additional variation in seasonal solar heating due to its eccentric orbit. In 1989, when Pluto was at perihelion (the nearest point to the Sun in its orbit), the light falling on it was three times stronger than it will be when Pluto is at the farthest point in its orbit, in 2113. The extreme seasonal variation of the position of the Sun in the skies of Triton and Pluto, and the strength of the sunlight falling on Pluto, suggest that the distribution of nitrogen ice on their surfaces is likely to be highly variable as that ice sublimes, is transported, and re-condenses in response to the changing sunlight. Consequently, the pressure of their atmospheres may also change considerably with season. For Pluto, it has been suggested that the huge decrease in the strength of sunlight with orbital distance will cause the nitrogen atmosphere to literally freeze-out completely on to the surface within a few decades.

Unfortunately the Voyager observations of Triton did not really tell us where the nitrogen ice was on the surface. The most prominent feature visible in images of Triton (Fig. 6.1) is the large, high albedo "South Polar Cap" (SPC). Three lines of evidence suggest that the SPC is composed of nitrogen ice, with small amounts of methane and carbon monoxide ice mixed in. First, most of the surface of Triton visible from Earth must be covered in nitrogen ice; otherwise the nitrogen absorption feature in Triton's near-infrared spectrum (Fig. 6.4) would be much weaker than it is. The SPC covers most of Triton's surface as viewed from Earth, so it is very difficult to explain the strength of the nitrogen feature unless the SPC is composed of nitrogen. Second, nitrogen ice preferentially sublimes from areas absorbing more than average solar energy, and condenses in areas where the absorbed solar energy is less than average. The SPC in some areas has such a high albedo (that is, those areas absorb so little sunlight) that nitrogen is probably condensing on them even though it is summer in Triton's southern hemisphere. This at least suggests a link between high albedo areas and the presence of nitrogen ice, although it does not require that all high albedo areas be covered in nitrogen. Third, the pressure of Triton's atmosphere was measured by Voyager, and by subsequent stellar occultations similar to the stellar occultation by Pluto. Triton's atmospheric pressure requires that the nitrogen ice on the surface has a temperature between 37 and 39 K. As discussed earlier, the temperature of the nitrogen ice depends on its global distribution. More specifically, Triton must have about half or more of its nitrogen ice in the southern hemisphere, where the Sun is shining, for the global ice temperature to be 38 ± 1 K. If the well-illuminated SPC is not dominated by nitrogen ice, there is little area left where nitrogen ice on Triton can absorb sunlight and warm itself, even to the frigid temperature of 38 K.

The fate of Pluto's atmosphere

Pluto's atmosphere may be ephemeral. Currently Pluto is near perihelion, the point in its orbit where it is nearest the Sun. The nitrogen ice on Pluto is probably about as warm as it ever gets; consequently the atmospheric pressure is probably about as high as it gets. As mentioned above, the intensity of the sunlight falling on Pluto varies by a factor of three during its year. As it continues to recede from the Sun over the coming decades its temperature and atmospheric pressure will tend to decline. The vapor pressure of nitrogen is a very strong function of temperature. For a 2 K change in temperature, Pluto's (or Triton's) atmospheric pressure will change by a factor of almost ten. Pluto's atmospheric pressure in 1989, at the time of the stellar occultation, was between about 10 and 40 microbars (10–40 millionths the pressure of Earth's atmosphere). Even at that low pressure the atmosphere really is an atmosphere. It is distributed quite uniformly around the globe, with moderate winds transporting subliming nitrogen ice between regions. The only other comparably sized bodies with "real" atmospheres in our solar system are Triton and Saturn's moon, Titan. All other moons and small bodies have, at most, atmospheres so thin that the surface pressure is not even really defined. On such objects the molecules of gas in the "atmosphere" are so few that they are as likely to escape from the surface into space as they are to collide with one another. Near the surfaces of Pluto and Triton atmospheric molecules travel only a centimeter or so before colliding with another gas molecule, so the atmospheres are very much in the fluid regime, as opposed to the molecular regime like the atmospheres of the Moon, Mercury, or the galilean moons of Jupiter.

When nitrogen ice in one area sublimes, the gas created is transported by winds across the surface to the region where it condenses. The rate at which the atmosphere transports gas from regions of sublimation to regions of deposition is proportional to both the atmospheric density and the wind speed. If the atmospheric density decreases the wind speed must increase to support the transport of nitrogen across the surface. If Pluto's (or Triton's) atmospheric pressure falls enough, it will become so thin that winds will begin to blow at speeds approaching the speed of sound. This would be the case if Pluto's atmospheric pressure fell by a factor of about 200, to around 0.1 microbars. Such high-speed winds, driven by sublimation and condensation of nitrogen ice, would require huge pressure differences across the surface: there would no longer be a well-defined surface pressure. In that case it would be fair to say that the character of the atmosphere was distinctly different from what it is at present. Also, because differences in pressure are equivalent to differences in temperature in Pluto's vapor-pressure atmosphere, the nitrogen ice in subliming areas would be warmer than that in areas of condensation. The nature of the interaction between surface nitrogen ice and the atmosphere would be transitional between the current dense atmosphere scenario and a situation where Pluto no longer has a "real" atmosphere. If Pluto's atmospheric pressure falls even further, well below 0.1 microbars, the atmosphere will basically have frozen-out on the surface. Any atmosphere left would be local, and subject to direct escape of the molecules into space.

Is it inevitable that Pluto's atmosphere will disappear in the next few decades as Pluto recedes from the Sun? The answer to this basic question is turning out to be particularly elusive because of the properties of nitrogen ice, the ability of that ice to move across the surface, and the complex seasonal forcing resulting from Pluto's eccentric orbit and tilt.

Nitrogen ice can exist in three different forms, called alpha (α), beta (β), and gamma (γ) phases. Nitrogen is in the α phase at temperatures below 35.6 K, and it exists in the β phase at temperatures higher than 35.6 K: currently the nitrogen ice on Pluto and Triton is β-nitrogen. The γ phase of nitrogen ice exists at very high pressures, and so is not relevant to the surface of Pluto (or Triton). Fans of Kurt Vonnegut novels may realize that water ice also has different phases, although ice-9, described in one of his books, is purely fictional. The difference between α- and β-nitrogen lies in their crystal structure. β-nitrogen is a somewhat disordered crystal, while α-nitrogen has a more ordered crystal structure. Just as there is a latent heat associated with changing from the ice phase to the gas phase, there is also latent heat associated with the nitrogen $\alpha - \beta$ phase transition. When β-nitrogen changes into α-nitrogen some heat is given off, and if the conversion goes from

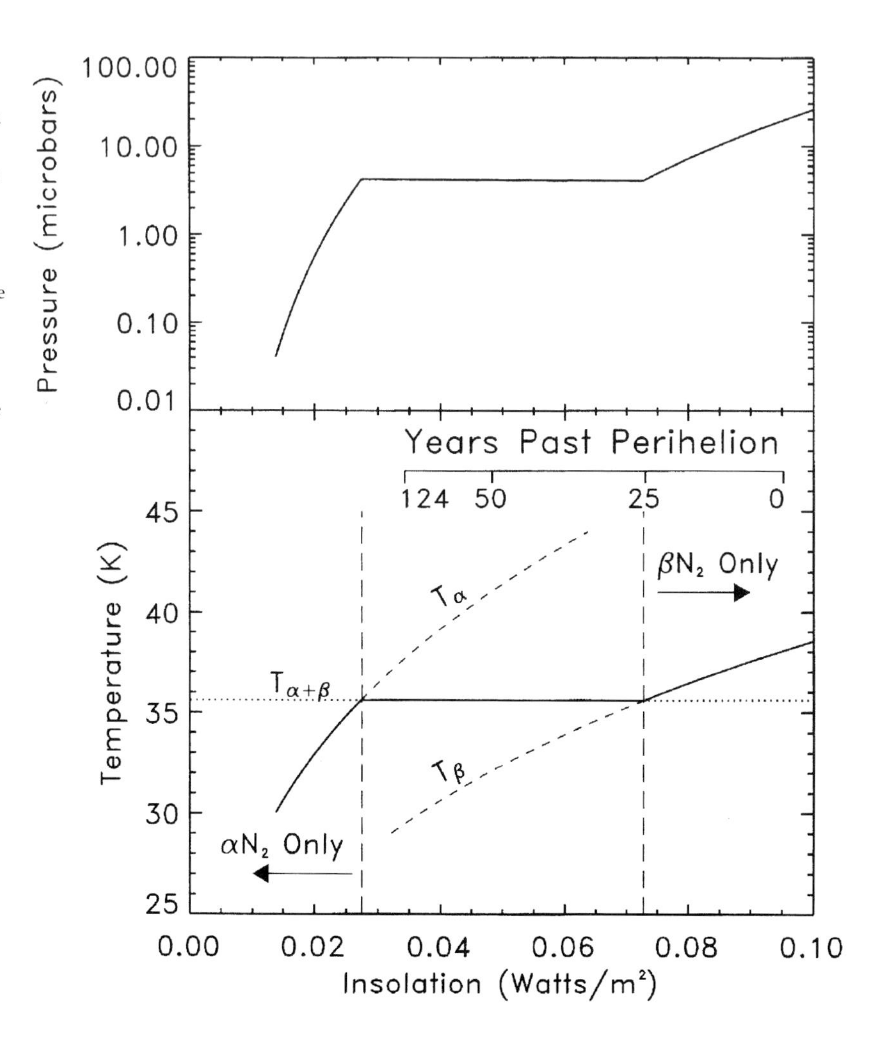

Figure 6.7 The temperature (lower panel) of Pluto's nitrogen ice as a function of the strength of sunlight falling on Pluto and/or the number of years past perihelion. At a temperature of 35.6 K the temperature of the nitrogen is buffered by a solid state phase change and a change in the emissivity of the ice that is accompanying the phase change. This simple model predicts that Pluto's atmosphere will never "freeze out" on the surface.
Credit: John Stansberry.

α to β, heat is absorbed. Once the surface of Pluto cools to 35.6 K, β-nitrogen will begin to be converted to α-nitrogen, and heat will be produced as a result. This release of latent heat will warm the surface, delaying atmospheric freeze-out.

Figure 6.7 shows the predicted temperature of Pluto's nitrogen ice, and pressure of its atmosphere as a function of time since perihelion in 1989. At perihelion the nitrogen ice is in the β phase, the temperature is 38 K, and the atmospheric pressure is 20 μbars. As Pluto moves outward in its orbit the intensity of sunlight decreases, and the temperature of the nitrogen ice declines until, at 25 years past perihelion, the temperature falls to the $\alpha - \beta$ phase transition temperature, 35.6 K, and the atmospheric pressure is 4 μbars. As Pluto moves farther from the Sun nitrogen ice begins to be converted from the α to the β phase. This phase transformation, like sublimation and condensation, is location specific: α-nitrogen forms in areas of little or no sunlight, releasing latent heat in those regions. However, the effect of the latent heat release is global because the global balance between absorbed insolation and thermal re-radiation determines the temperature of the ice. Assuming that a globally distributed layer of nitrogen ice about 1 meter deep must undergo the $\beta - \alpha$ transition before temperatures can drop further, atmospheric freeze-out is delayed about one decade by this latent heat effect.

There is a subtle but important distinction between a Pluto with only β-nitrogen ice, and a Pluto with both α- and β-nitrogen. The atmospheric pressure for an ice temperature of 35.6 K is 4 μbars, well within the range where atmospheric energy transport keeps all of Pluto's nitrogen ice at a single temperature. So long as both α and β nitrogen are present on the surface the atmospheric pressure has to be 4 μbars. This is very important if the two phases differ in a physical characteristic which affects their temperature, for example their emissivity. A material with a low emissivity is inefficient at cooling by radiating heat away, so for a given amount of energy input (e.g. sunlight) a low emissivity material will be warmer than a high emissivity material. Emissivity is very difficult to measure for a substance like nitrogen ice because it readily sublimes and can condense out on the optical equipment being used to measure the emissivity. The emissivity of nitrogen ice has never been measured, but its far-infrared absorption spectrum has. Emission and absorption of photons are physically symmetric processes, so knowledge of the absorption spectrum of nitrogen can be used to calculate its emissivity.

The heat radiated by Pluto peaks at wavelengths of 70–100 μm, 100 times longer than the wavelengths of visible light. Nitrogen ice is a very weak absorber at those wavelengths, just as it is a weak absorber at near-infrared wavelengths (Fig. 6.4). As a result, it is an inefficient emitter of thermal radiation. β-nitrogen has a broad, weak absorption band at 140 μm, giving it a moderately low emissivity, probably around 0.8 (emissivity, like albedo, ranges between 0 and 1). The crystal structure of α-nitrogen is much more ordered than that of β-nitrogen, and as a result it has very narrow absorption bands, at 140 and 204 μm. These bands are also quite weak, and because they are very narrow, α-nitrogen on Pluto (and Triton) has an emissivity of about 0.25.

Figure 6.7 illustrates the effect of this emissivity contrast on Pluto's seasonal evolution. About 25 years after perihelion, Pluto's nitrogen ice cools to the $\beta - \alpha$ transition temperature, and α-nitrogen begins to form in un-illuminated areas, while β-nitrogen is the phase in illuminated areas. If there were no such thing as α-nitrogen Pluto's temperature would continue to decline along the dashed curve extending down and to the left from the 35.6 K, 25 years past perihelion point; with α-nitrogen the temperature instead moves horizontally along the solid line. This is because the temperature of the ice is determined by the balance between sunlight absorbed by the β-nitrogen and thermal emission by all of the nitrogen. The α-nitrogen has a lower emissivity than the β, so the emissivity of the whole is somewhat lower than it would be if only β-nitrogen were present, and the temperature correspondingly higher than it would be in the all β case.

As Pluto moves farther outward more ice is converted from the β to the α phase, the temperature remains at 35.6 K (as it must for the atmosphere to be in vapor–pressure equilibrium with the surface ices), and the atmospheric pressure remains at 4 μbars. The emissivity contrast between the two phases of nitrogen ice determines when all of the nitrogen is converted to the α phase, and when the atmospheric pressure will begin to decline again. If the emissivities of α- and β-nitrogen are 0.25 and 0.8, there is still some β-nitrogen left on Pluto's surface at aphelion, and the atmosphere is just as dense as it was only 25 years past perihelion. The emissivity calculations are somewhat uncertain, and the emissivity of α-nitrogen might be as high as 0.5. In that case it would take 70 years after perihelion for all of the nitrogen to be converted to the α phase. At that time the atmospheric pressure would begin to drop again, but atmospheric freeze-out would have been postponed by 45 years.

The simple models described so far provide some indications that Pluto's atmosphere may not be doomed to freeze-out on the surface soon, or even ever. In August 2002 Pluto occulted a star that was bright enough to provide another opportunity to study the atmosphere. Pluto's star-shadow passed over the observatories on the Hawaiian islands and in southern California, and very high-quality data were obtained at both sites, as well as a few others. The new stellar occultation showed that, contrary to the predictions of simple models, Pluto's atmospheric pressure is actually higher now than it was in 1989, by a factor of two or three. This increase, while somewhat unexpected, is not a big surprise because the surface–atmosphere system of Pluto is complex and our knowledge of its detailed state is very poor. No doubt Pluto has many more surprises in store.

Not yet explored

As of this writing, NASA is preparing the first exploratory mission to Pluto: the New Horizons mission. Slated for launch in 2006, New Horizons will swing by Jupiter for a gravity assist, and will require about 12 years to reach Pluto. Outfitted with two visible imagers, a near-infrared mapping spectrometer, ultraviolet and radio occultation experiments, and a dust detector, New Horizons will provide a snapshot in time of conditions on Pluto's icy surface and in its atmosphere in 2018. We will then have better answers to the question of whether Pluto is really a twin to Triton, or a completely unique body. Will there be an atmosphere for the spacecraft to study? Will the surface be heavily cratered, or young like Triton's? After the Pluto encounter, New Horizons will continue outward, encountering one or more trans-Neptunian objects 2–3 years later, and providing our first detailed look at members of that population, and doubtless raising more questions about them than it actually answers . . .

DALE P. CRUIKSHANK
NASA Ames Research Center

Comets: ices from the beginning of time

Comets – They are messengers from beyond the planets, carrying material unchanged from the earliest times in our galaxy, and bringing to Earth both the matter of life and the violent means to extinguish it. Comets are the givers and the takers of life. As our understanding of the nature of comets has surged ahead in recent years, we have come to recognize that these objects have an extraordinary story to tell about the earliest days of the solar system, at a time when planets were just forming and the Sun was a dimly glowing ember against the darkness of space.

Each time we take a drink of water or inhale a breath of air, each time we gaze at the grass and trees, and each time we greet one of our fellow humans or another member of the animal kingdom, we are encountering matter that ultimately came to Earth from comets and asteroids. Astonishing as it may seem, a significant amount of the volatile material on Earth, including water, the nitrogen of the air, and the carbon of the biosphere, arrived on our planet early in its history, and very soon after it had begun to cool from the initial heat of accumulation. Experts still debate the quantities of extraterrestrial volatile material delivered to the early Earth, but almost all agree that the amounts were substantial.

The impacts of countless icy comets and carbon-rich asteroids in the first half billion years of Earth's history profoundly affected this planet, fracturing and stirring the crust, and delivering the stuff of life. After life somehow sprung forth, later impacts caused global ecological changes and altered the course of biological evolution, perhaps many times. A cosmic impact may be implicated in the greatest mass extinction of species in the fossil record of life on Earth that occurred about 250 million years ago toward the end of the Permian period. Further, scientists generally agree that the global catastrophe 65 million years ago, which caused the extinction of the dinosaurs and the majority of all animal species on the Earth, was caused by the impact of a comet or an asteroid.

What are these comets that have had such a profound effect on the Earth, and where do they come from?

Figure 7.1 Comet Hyakutake, the Great Comet of 1996, was discovered by Yuji Hyakutake on January 30 of that year. It made its closest approach to Earth on March 25, at a distance of 9.3 million miles, and was closest to the Sun on May 1. This photo shows the thin strands of the ion tail encased in the diffuse dust tail.
Credit: photo by Loke Kun Tan © StarryScapes.

Cloaked in mystery and superstition for millennia, bright comets have always evoked a sense of excitement and foreboding. While the heavens were thought to work in an orderly fashion, with the Sun and the Moon, the stars and planets moving predictably with clockwork-like precision, the sudden appearance of a comet upset the picture. As often as not, momentous events seemed to occur with the appearance of a bright comet – a royal birth, a decisive battle, the death of a monarch, or some natural catastrophe. It was easy for superstitious people to attribute notable events such as these to the comet itself. Now we know better (or do we?)

In earlier years, comets often seemed to appear from nowhere, showing in the evening or morning sky as a fuzzy patch of light with a tail streaming outward in a direction opposite the Sun. Now, many comets are found by avid amateur astronomers who search the skies with small telescopes, or by automated camera-telescopes looking for asteroids or comets that may someday imperil the Earth. Comets are most often discovered before they develop into anything visible to the naked eye.

In fact, the vast majority of comets never become bright enough to be seen with anything but a fairly large telescope (Fig. 7.1).

The last century of telescopic studies of comets and visits of five spacecraft to the most famous one, Comet Halley, have revealed many, but not all of the secrets of the comets. More spacecraft are on their way to comets, and still more are in the planning stages (see below).

What are comets?

As a comet's orbit brings it closer and closer to the warmth of the Sun, it becomes larger and brighter, developing a fuzzy appearance (the coma) and a prominent tail that can extend more than 150 million kilometers in length (more than the distance from Earth to the Sun). After the comet's closest approach to the Sun, it begins to head back to the cold outer reaches of the solar system. Eventually it becomes fainter, the tail fades, and the comet finally becomes invisible. The time scale for this cycle can be anything from many months to a few days, depending mostly on the distance of the comet from the Sun (and Earth).

Recognizing this pattern of behavior, in 1950 the US astronomer Fred Whipple promoted his concept of comets as "dirty snowballs." The "dirt" is now known to consist of dusty particles of minerals and solid organic matter. The "snow" is understood to consist of frozen water and numerous other volatile materials that make up the overall ice component that is so important in governing the comet's behavior as it alternately warms and cools along its orbital path around the Sun. The heat of the Sun causes the ices to evaporate and to escape from the solid body of the comet (the nucleus). As the gases from the evaporating ices seep out through the comet's surface, they carry along small particles of the dusty material that form the large but very thin coma surrounding the nucleus. The coma appears bright because the dust is illuminated directly by visible sunlight, and because some of the gases are caused to glow (fluoresce) by the action of the ultraviolet light of the Sun.

Whipple also recognized that a typical comet nucleus is very small on the cosmic scale of things – most often about one or two kilometers in diameter. There are a few larger ones, but in the past century the largest comet to pass by the Earth was probably Comet Hale-Bopp at about 40 kilometers in size (Fig. 7.2).

Frozen water is one of the most abundant of all materials in the colder regions of space, both in the solar system and in the regions

Figure 7.2 The spectacular Comet Hale-Bopp lingered in the sky for several months in 1997, appearing at its brightest in April. Its closest distance to Earth approached 200 million kilometers.
Credit: photo by Loke Kun Tan © StarryScapes.

between the stars throughout the galaxy. Interstellar ices (and dust) are in fact the progenitors of the solid material in the solar system. The full story of interstellar matter is outside the scope of this chapter, but the connection to comets is too important to pass by. Let us therefore give a brief description of the origin of interstellar dust and ices.

The interstellar medium, and the death of stars

When stars of various kinds approach the ends of their "lives," they often eject large quantities of dust grains composed of silicon-rich molecules in the form of olivine and pyroxene minerals (combinations of silicon and oxygen plus iron and other metals) and silicon carbide (silicon plus carbon). From some stars, carbon-rich grains already containing complex organic molecules (combinations of carbon, hydrogen, oxygen, and nitrogen) are ejected. These liberated materials are shot into space, and those that leave the gravitational confines of the parent star are left to mingle in interstellar space, enriching the interstellar medium. Mineral grains of silicon floating in the cold regions of interstellar space, where vast molecular clouds have accumulated, are slowly coated with molecules of water, carbon dioxide, carbon monoxide, and other volatiles, which eventually build up a layer of ice. The ice coating may be only a few molecules thick, but when ultraviolet light from a nearby star strikes it, chemical reactions can take place. The general effect is to produce more complex molecules from the simple molecules that originally condensed on the grain, and the result can be the formation of a solid organic residue that is not easily evaporated (refractory). These "macromolecular" organic materials can be complex masses of large molecules, perhaps a little like coal or crude oil, although they do not have a biological origin as do coal and oil on Earth.

Thus, we have in interstellar clouds the main constituents of comets: ice, silicate dust, and large organic molecules. But how do interstellar clouds make comets?

New stars form when local "lumps," or condensations occur in giant interstellar clouds of gas and dust. The gentle, but inexorable force of gravity pulls gas and tiny dust grains together, and in the right circumstances, rotating, collapsing masses of gas and dust quickly form stars. One of the important stages in the formation of a star is the development of a dense disk of gas and dust around the central condensation, and it is in the disk that planets and smaller solid bodies (such as comets) originate. Thus, the icy grains from the interstellar medium are important constituents of any solid bodies that might form around stars.

Comets are formed

The rotating disk of gas and dust that gave birth to our own Sun is called the solar nebula. As the Sun was gathering its mass and preparing to ignite its internal nuclear furnace, condensations and collisions of dust particles in the outer regions of the nebula were gradually building planetesimals, the macro-scale objects from which the planets were eventually built. Comets are planetesimals that were never incorporated into large planets. They began as individual icy grains collided and stuck together in the process called "accretion." Accretion occurs on a familiar scale under furniture and in unswept corners of a room, as flimsy "dust bunnies" appear and grow through the accumulation and sticking together of invisible particles of dust, lint, and other household debris. Comets began as the icy dust bunnies of the solar nebula.

The detection of various atoms and molecules in comets, using optical, infrared, and radio telescopes, and the comparison of the abundances of those materials with the abundances observed in the interstellar gas, supports the contention that the basic interstellar abundances are preserved through the process of comet formation in the solar nebula. In particular, the abundances of the deuterated versions of water (HDO instead of H_2O) and of hydrogen cyanide (DCN instead of HCN) in Comet Hale-Bopp are in the same ratio as they are in the interstellar gas. Deuterium is an isotope of hydrogen that contains a neutron in addition to the proton and electron of an ordinary H atom; most of the deuterium was originally formed in the Big Bang that created the universe.

At the same time, comets are not simple, structureless aggregates of interstellar grains. As planetary scientist Stuart Weidenschilling has pointed out, comets display complex outbursts and jetting of gas that suggest compositional inhomogeneities on a scale of 20–100 meters. The lumpy nature of comets probably arises from the accretion of kilometer-size or larger nuclei from smaller planetesimals that formed at different times and/or different locations in the solar nebula. Also, it is possible, perhaps likely, that chemical changes occurred in some of the materials of the solar nebula, particularly in regions of high density and high temperature. The degree to which such chemical changes took place, and just how much of that altered material is incorporated in the comets, is unknown. Experts are particularly intent on exploring this issue, but they generally agree that much of the material in comets is original interstellar matter (Fig. 7.3).

The composition of comets

The astronomer's standard technique for determining the composition of a celestial body is spectroscopy. In the case of a comet, when it is close enough to Earth to study by spectroscopy, it is also so close to

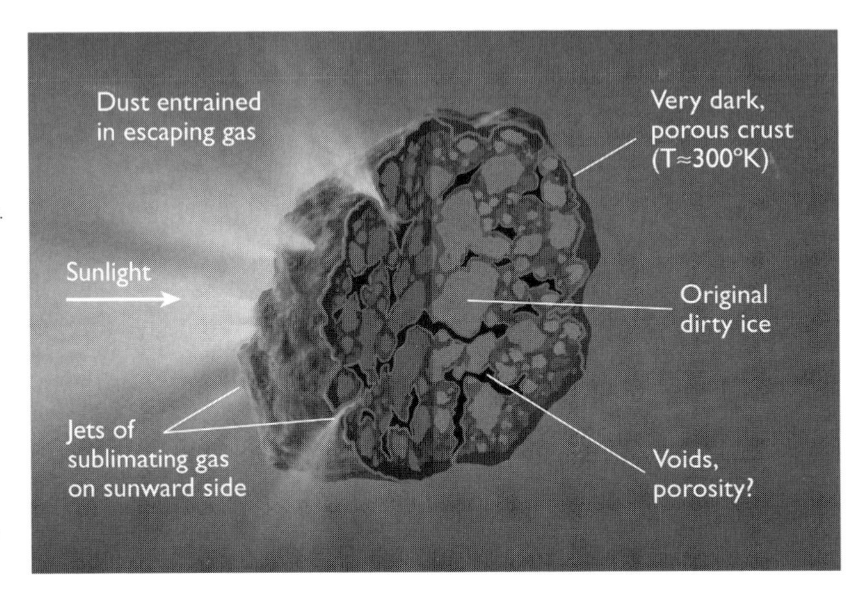

Figure 7.3 The nucleus of a comet is a loosely compacted lump of ices, silicate mineral dust, and complex organic matter similar to soot. As the comet approaches the Sun, the warming ices begin to evaporate. As the gas seeps through the porous nucleus, it carries along minute particles of the mineral dust and organic matter, discharging them into space to form the comet's coma and tail. Some parts of the nucleus are "active," with jets of dust-laden gas erupting from the surface, while other parts are inactive. The level of activity increases as the heat of the Sun penetrates into the nucleus, and then diminishes as the comet recedes from the Sun on its elongated orbit.
Credit: Cometary nucleus by Don Davis from *The New Solar System*, 4th edn., edited by J. Kelly Beatty, Carolyn Collins Petersen, and Andrew Chaikin, Sky Publishing, 1999.

the Sun that the solid nucleus is enshrouded in the opaque coma produced by dust and the evaporating ices. Therefore, we cannot usually observe the solid surface of a comet directly, but we are making significant progress in learning the compositions of the comets through some indirect techniques. The most informative indirect method is to study the escaping gases themselves, and then try to understand which kinds of ice evaporated to produce them. This is complicated in some instances by the fact that chemical reactions occur between the time the ice evaporates and the gas is observable with a spectrometer.

Another problem that has vexed comet specialists is the observation that gas appears to be coming not only from the solid nucleus, but simultaneously from a larger source, the coma itself. As gas leaves the nucleus, it carries solid particles of comet material, and then additional gas is emitted by these individual particles. The particles consist in part of ice, the evaporation of which produces gas, but the particles also consist of organic solids mentioned earlier, and these organic molecules may be decomposing to produce even more gas. For a given comet, the relative amount of gas evaporated from the nucleus and from particles in the coma changes dramatically as the comet's temperature changes. Comet Hale-Bopp, in particular, produced carbon monoxide gas both

Table 7.1 *Molecules found in comets*

Symbol	Name	Symbol	Name
H_2O	Water	HCN	Hydrogen cyanide
OH	Hydroxyl radical	HNC	Hydrogen isocyanide
H_2O^+	Ionized water	HNCO	Isocyanic acid
CO	Carbon monoxide	CN	Cyanogen
CO_2	Carbon dioxide	N_2^+	Nitrogen ion
CO^+	Carbon monoxide ion	NH	Imidyl radical
HCO^+	Formyl ion	NH_3	Ammonia
C_2	Carbon (diatomic)	NH_2CHO	Formamide
C_3	Carbon (triatomic)	CH_4	Methane
H_2S	Hydrogen sulfide	C_2H_2	Acetylene
OCS	Carbonyl sulfide	C_2H_6	Ethane
CS	Carbon monosulfide	HC_3N	Cyanoacetylene
CS_2	Carbon disulfide	CH_3CN	Methyl cyanide
SO_2	Sulfur dioxide	H_2CO	Formaldehyde
SO	Sulfur monoxide	$HCOOCH_3$	Methyl formate
H_2CS	Thioformaldehyde	HCOOH	Formic acid
S_2	Disulfur	CH_3CHO	Acetaldehyde

Table 7.2 *Isotopic varieties of cometary molecules*

Symbol	Name	Symbol	Name
HDO	Deuterated water	DCN	Deuterated cyanide
$H^{13}CN$	–	$H^{15}N$	–
^{34}CS	–		

from the nucleus and from particles in the coma when the comet was less than about 1.5 AU from the Sun.

All the complications notwithstanding, we now know that the dominant ices in comets are H_2O, carbon monoxide (CO), carbon dioxide (CO_2), methyl alcohol (CH_3OH), and methane (CH_4). Numerous other molecules (see Tables 7.1 and 7.2 below) are also detected in small abundances, and these also come from cometary ices, as do the more abundant species.

Whipple's dirty snowball, or icy conglomerate model predicted that ordinary water ice is the most abundant. Indeed, subsequent observations have shown this to be true, although the first water in comets was positively detected only some 35 years after Whipple's prediction, when observational techniques had developed to the point that the measurement was possible in Comet Halley by astronomer Michael Mumma and his colleagues. Some comets evaporate great quantities of water ice when they are near the Sun; Comet Hale-Bopp emitted 300 tons of water vapor per second during its passage through perihelion, even though it did not get as close to the Sun (0.914 AU = 1.37 million kilometers) as many comets do. This quantity is about the amount of water that a large swimming pool holds. At the same time, Comet Hale-Bopp was discharging 100 tons of CO per second.

While many of the molecules in comet ices contain hydrogen atoms (e.g., H_2O, CH_3OH, and H_2CO), others, such as CO_2, CO, and SO_2, do not. Because of the special chemical properties of hydrogen, in particular its ability to bond with other atoms, this distinction is an important indication of the earliest history of the ices. Recent studies of interstellar ices show the same chemical distinctions; "protic" ices with hydrogen and "non-protic" ices without. We can use these properties of different ices to help determine in just what kind of regions of interstellar space the materials of comets formed. That is, what were the chemical environments, in terms of temperature and the availability of materials?

The two fundamentally different kinds of ice appear to originate in regions of interstellar space where the hydrogen occurs mostly in the form of atoms (H) or the hydrogen molecule (H_2). In regions in the interstellar medium where the H/H_2 ratio is large, H atom addition (hydrogenation) is the dominant chemical process, and species such as CH_4, NH_3, and H_2O are expected to be prominent. One such example is the line of sight towards protostar W33A, where water and methyl alcohol (both containing H) are clearly seen. Laboratory studies by chemist Max Bernstein and others, of ice mixtures containing water, methanol, and ammonia irradiated by ultraviolet light (UV photolysis) show that complex organic molecules readily form from these simple starting materials.

Ices from environments rich in molecular hydrogen, where H/H_2 is $\ll 1$, appear to produce oxygen-rich, non-protic ices. One such

H_2-dominant region in the galaxy is consistent with what is seen in the molecular cloud toward the star Elias 18, where spectroscopy indicates little water in the ice and perhaps considerable amounts of molecular oxygen, or nitrogen, or both.

If material from both kinds of interstellar environments was mixed in the giant molecular cloud in which the solar system was born, perhaps it is not surprising that comets are heterogeneous mixtures of ices and dust of different kinds. Nor is it surprising that there can be irregular and episodic ejections of gas, as various kinds of ice are warmed by the Sun to the point of evaporation.

Special properties of water ice

Water ice forms on an interstellar grain at the exceedingly low temperature of typically 10 to 25 K (−263 to −248 C), depending on the local environment and the distance from the nearest stars. At such a temperature the ice forms in an amorphous state, with no definite crystalline structure of the kind seen in snowflakes or in patches of ice on a window pane. If the ice is heated to about 140 K (−133 C), it converts spontaneously to a crystalline structure, first cubic, and then hexagonal. This conversion is not reversible, even if the temperature is lowered drastically, unless the crystalline structure is broken by the impact of an atomic particle, such as a cosmic ray. Local heating of an icy surface, by such effects as meteoroid impact, can also convert amorphous ice to crystalline, even if the average temperature of the surface is less than the conversion temperature of about 140 K.

Because comets formed at exceedingly low temperatures, their water originally froze in the amorphous state. But when a comet approaches the Sun for the first time, its ice begins to convert to crystalline by about the time it reaches 5 AU, or the distance of Jupiter from the Sun. This change occurs primarily on the comet's surface, but the conversion to the crystalline phase releases a quantity of heat that propagates inward to the deeper layers of the comet, causing ice there to convert from amorphous to crystalline, as well. The depth of the penetration is thought to be some 25 to 40 meters, but it depends on the orbit of the comet and the nature of its icy surface layer.

Another property of water in comets that supports the idea that it was frozen solid at the very low temperatures characteristic of the interstellar medium, rather than in the warmer regions of the solar nebula, is the direction of spin of the nuclei of the two hydrogen atoms of the water molecule. In ortho water, the two H atoms spin in the same direction, and in para water, they spin in opposite directions to one another. The two kinds of spin can be distinguished spectroscopically, and the ratio of the quantity of ortho water to para water is an indication of the temperature at which the molecules were frozen into ice. Observations of the ortho/para ratio in Comet Halley was found by Michael Mumma and his colleagues to be 2.5 ± 0.1, which suggests that the ices of this comet froze at a temperature of about 29 K. Similar measurements of Comet Hale-Bopp give a freezing temperature of about 25 K; both are consistent with the freezing of water at the temperature of the interstellar medium, rather than the warmer temperatures thought to have existed in the solar nebula.

Still another effect of the very low temperature of formation of comets is that their ices trap various quantities of different gases present in the local environment. Of special interest are the noble gases (gases that do not interact chemically with other materials) Ar (argon), Kr (krypton), and Xe (xenon). If the temperature of ice formation is 30 K, these gases are trapped in the original proportions in which they occur in the interstellar medium. However, at a slightly higher temperature of 50 K, Ar is trapped much more poorly than Kr and Xe, and the proportions of these atoms in the ice are therefore different from the proportions in the surrounding interstellar gas. In principle, then, we can determine the temperature of formation of a comet's ices from the proportions of various gases, including the noble gases, observed in the coma and tail as the comet passes near the Sun. Using this basic idea, planetary scientist T. C. Owen has estimated the amount of gas of various kinds in the atmospheres of Venus, Earth, and Mars that were brought to those planets by the impacts of comets (see Chapter 3 of this book).

What comets are made of

Spectroscopy of comets extends across a huge range of wavelengths, from the far ultraviolet to the radio region, including those wavelength regions observable only from space. Important compositional information is found at all wavelengths, with radio telescopic observations growing in importance because of their great sensitivity to very small quantities of many different kinds of molecules.

A tremendous advance in our knowledge of the chemical composition of comets came from the fortuitous appearance of two exceptional comets. Comet C/1996 B2 (Hyakutake) approached the Earth at a distance of only 0.1 AU in March 1996, and was studied intensively with newly developed infrared and radio astronomical instruments. Then, comet C/1995 O1 (Hale-Bopp) reached perihelion in March–April 1997, and was very bright, making it well suited to detailed infrared and radio observations.

The gases of comets, including many new ones found in comets Hyakutake and Hale-Bopp, are shown in Table 7.1. Many of these are not seen in any other solar system bodies, but most of them have been seen in interstellar clouds.

Table 7.1 does not tell the whole story, however, because certain of these molecules are found in different isotopic forms. That is, one of their atoms is an isotope that is different from the normal, most abundant form. Table 7.2 shows the isotopically different molecules detected in comets.

Of particular importance is water with one deuterium atom, as noted above. The ratio of HDO to H_2O in comets Halley, Hyakutake, and Hale-Bopp is about two times that of the water on Earth, and about ten times that in the solar nebula (determined from the composition of the Sun and certain kinds of meteorites). Knowledge of this fact allows us to conclude two important things: first, only about 15% of Earth's water could have come from comets of this kind, otherwise the deuterium in our oceans would be more abundant than it is. Second, the deuterium content of comets provides further confirmation that they consist of materials frozen initially in the interstellar cloud, and were not reconstituted in the warmer solar nebula from which the Sun and planets formed. That is, they are truly primitive material that has not suffered the chemical effects of heating.

Comet dust

In addition to the large volume of water produced by Comet Hale-Bopp, a tremendous amount of dust, in the form of clumps up to several millimeters in size, was ejected as the comet passed the Sun. Astronomers

David Jewitt and Henry Matthews observed the heat from these particles, consisting of silicate and organic dust probably mixed with ices, and calculated that at perihelion the comet was losing more than 2,000 metric tons per second. The dust loss rate amounts to more than five times the loss rate of water (in the form of evaporated gas). In all, Comet Hale-Bopp lost about 3×10^{13} kilograms of material in the form of large dust particles during its passage by the Sun in 1995. This seems like a lot, but for Comet Hale-Bopp, which is about 40 kilometers in diameter, the loss of dust during this one passage by the Sun amounts to only about 0.1% of the total mass of the comet.

Dust produces much of the spectacular display of a big, nearby comet, forming a large, bright coma and stretching out in a long tail. We have several means of learning about the composition of the tiny flecks of comet dust, much of which is microscopic in size. First, we can tell from its spectroscopic properties that much of the dust is made of the minerals olivine [$(Fe, Mg)_2 SiO_4$] and pyroxene [$(Fe, Mg, Ca) SiO_3$]. Both of these minerals are combinations of Si (silicon), O (oxygen), and the metals Fe (iron), Mg (magnesium), and Ca (calcium), and both are formed in the outflow of gases from M-type giant stars, which are much cooler (but very much larger) than our Sun. The same minerals, formed in different ways, are common in Earth's volcanic rocks, and in meteorites.

Second, we can collect tiny particles of dust from comets here on Earth, or most accurately, from high in the Earth's atmosphere. Passing comets litter the inner solar system with their dust, and this dust is intercepted by Earth and the other planets as we orbit the Sun. Comet dust entering the atmosphere usually burns up in flashes of light that we see as meteors, or "shooting stars" in the nighttime sky. However, some of the tiniest particles weighing about a billionth of a gram enter the atmosphere more gently, and float in the high stratosphere for many weeks before falling slowly to the ground. Research aircraft capable of flying at 18,000–24,000 meters altitude (60,000 to 80,000 feet) have been collecting these microscopic comet dust particles for several years, and some 10,000 of them have been identified and studied. Even though their combined weight is much less than one gram, these interplanetary dust particles, or IDPs, have given us a remarkable amount of information about comets. Not all IDPs come from comets (some are fragments of asteroid collisions), but those of cometary origin contain

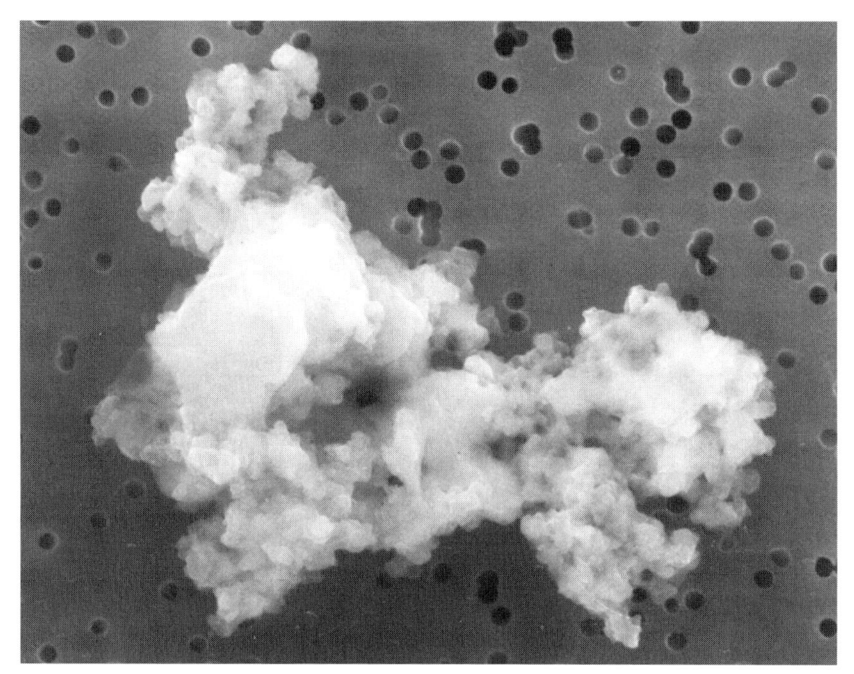

Figure 7.4 A particle of comet dust, after the ice has evaporated, is exceedingly tiny and very porous. This interplanetary dust particle (IDP) was captured in the stratosphere by a high-flying airplane, and was subsequently identified in the laboratory as a probable comet fragment. IDPs from comets contain silicate minerals, complex organic (carbon-based) chemicals surrounding the mineral grains, and ultra tiny flecks of iron and nickel metal. The IDP shown here is an aggregate of many grains that are 0.1 to 0.5 micrometers in size. The width of this entire particle is only 1/100 that of a human hair.
Credit: NASA/JPL/Caltech.

the silicate minerals noted above, plus an array of complex organic molecules (Fig. 7.4).

We have also learned about organic material in comets by the visits of the unmanned Giotto and Vega spacecraft to Comet Halley in 1985. Both spacecraft had special detectors aboard for the study of comet dust, and both found an astonishing array of complex organic molecular material in the dust of Halley. These particles, dubbed CHON because they consist of various combinations of C (carbon), H (hydrogen), O (oxygen), and N (nitrogen), together with the silicates discussed above, constitute the dust of comets.

Space scientists are greatly intrigued by comets and the stories they tell about the earliest times in the solar system, and, as a result, various space missions to comets are in progress or in the planning stages. In particular, the Stardust mission, launched in February 1999, promises to tell us a great deal about Comet Wild 2, because it has collected dust from that comet and will return it to Earth (in 2006) for detailed analysis. The progress of the Stardust spacecraft can be followed on the project's web site: http://stardust.jpl.nasa.gov (Fig. 7.5).

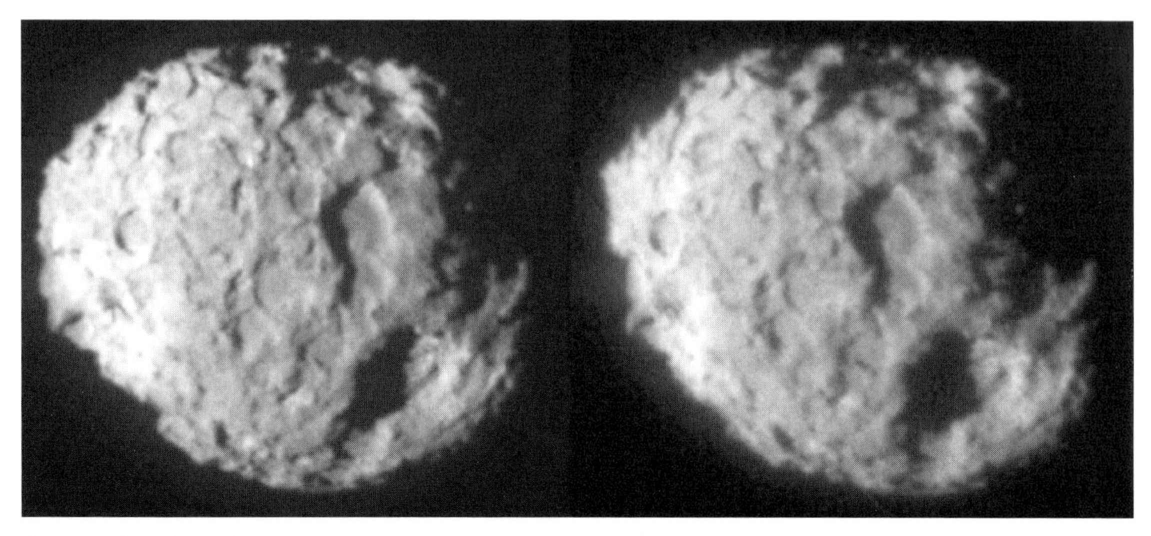

Figure 7.5 These views of the nucleus of a comet were taken during the close approach phase of the Stardust spacecraft's January, 2, 2004 flyby of comet Wild 2, which is about 5 km in diameter. These two images can be viewed as a stereoscopic pair. The surface shapes and features are unlike those seen on any other Solar System body. The Stardust spacecraft gathered dust from the comet, as well as other dust in interplanetary space, and will return it to Earth in January 2006 for analysis. This is the first space mission to bring back comet dust, and the information that scientists obtain from that dust will vastly expand our knowledge of the origin and evolution of comets, and their effects on the Earth and the other planets. *Credit*: NASA/JPL/Caltech.

Where do comets come from?

Comets making their first visit to the region of the Earth and other inner planets seem to come from two distinct regions of the solar system. Some come in on orbits that are highly inclined to the plane of the major planets' orbits around the Sun. They have very long periods of revolution around the Sun amounting to several hundred to several thousand years to a few million years. Other comets arrive on orbits of shorter period and are aligned closely to the plane of the planets' orbits.

Seeking to understand the source of the long period, highly inclined comets, astronomer Jan Oort made a study in 1950 and suggested that the solar system is surrounded by a halo, or sphere, of trillions of comets. The halo is so large that it extends almost half way to the nearest star (Proxima Centauri), which is 4.2 light years away. This reservoir, called the Oort Cloud, has never been seen directly, but it is thought to begin at about 20,000 AU and extend to more than 50,000 AU from the Sun. Each icy body in the Oort Cloud is in a huge orbit around the Sun. At the great distances of Oort Cloud bodies, the gravitational effects of nearby stars and the mass of the Milky Way Galaxy itself have stirred up their orbits, with the result that a complete range of inclinations is represented. If we could view a snapshot of the Oort Cloud, it would

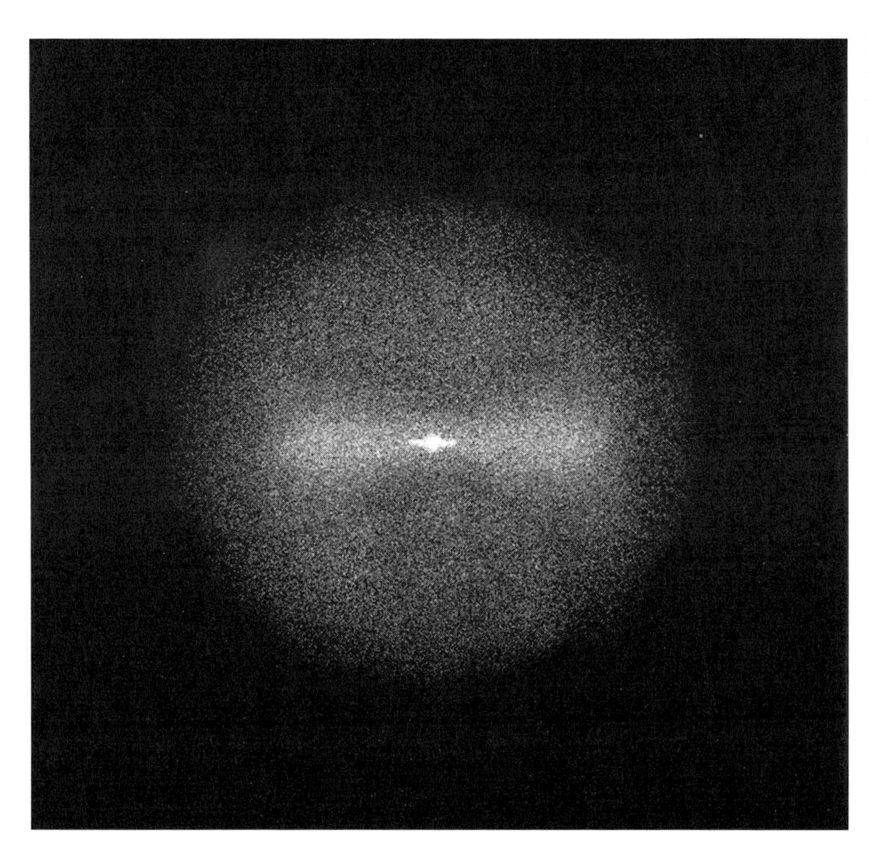

Figure 7.6 The Oort Cloud of distant comets contains trillions of icy bodies loosely held by the gravity of the Sun. Predicted by Dutch astronomer Jan Oort in 1950 as the source region of the long period (greater than ~200 years) comets, the spherical aggregate has a diameter of about 100,000 AU, and extends almost halfway to the nearest star. About half of the comets that approach the Sun every year, including comets Hyakatake and Hale-Bopp, come from the Oort Cloud, probably as a result of the gravitational perturbations of stars that slowly pass by the solar system. *Credit:* NASA Ames Research Center.

appear as a vast spherical cloud of icy bodies surrounding the Sun (Fig. 7.6).

The Oort Cloud is the source of about half of the comets that arrive in the inner solar system and become visible. At their usual distances far beyond the planets, they are so small and distant as to be invisible with our current telescopes. We see the occasional visitor from the Oort Cloud, and we can estimate the cloud's dimensions and the number of objects populating it, but we can not see it directly. Recent comets from the Oort Cloud include Comet Hyakutake, which made a spectacular appearance in early 1996, and Comet Hale-Bopp, which was one of the great comets of the twentieth century and was easily visible to the naked eye for several months in early 1997. Comet Halley, perhaps the most famous comet of all, was originally an Oort Cloud comet, as we can see from the fact that it now orbits the Sun in a retrograde orbit, going in the opposite direction to the planets themselves. Because

Comet Halley made a fortuitous close approach to Jupiter sometime in the distant past, it now travels in an inclined orbit with a period of about 76 years. Its next return to the vicinity of the Earth will be in the year 2061.

Every 30 to 40 million years or so, a star passes within about 10,000 AU of the Sun, causing a gravitational disruption of the Oort Cloud. This, and the less-frequent encounter of the Sun with a giant molecular cloud of the galaxy can occasionally dislodge a number of comets from the Oort Cloud. They can either leave the cloud entirely, or they can make a pass close to the Sun, where their orbits are often changed by interactions with the planets. There are other factors that send these distant icy objects into our region of the solar system from time to time, and there may even be comet "showers" that punctuate the steady, slow delivery process with pulses of comets arriving by the hundreds each year for a period of time.

The other comets that approach the Sun come from a different, closer, source. We know it is closer because the orbital periods of these comets are typically less than about 200 years. Furthermore, the comets approach in orbits that are near the same plane as those of the planets. In 1951, astronomer Gerard P. Kuiper proposed that these comets come from a source region just beyond Neptune, the outermost giant planet. He suggested that millions of kilometer-sized objects form a disk that lies in the plane of the planetary orbits, and contains the icy material left over from the formation of the planets. A similar suggestion had been made two years earlier by a less-prominent astronomer, K. E. Edgeworth, but the disk of short period, low-inclination comets is now generally called the Kuiper Belt or the Kuiper Disk. Just how far beyond Neptune the Kuiper Disk extends is unknown. It may continue far into space, perhaps eventually blending seamlessly into the Oort Cloud. Or, the Kuiper Disk may be more like a donut, extending only to 50 or 100 AU (Fig. 7.7).

Comets are "extracted" from the Kuiper Disk primarily by the gravitational action of Neptune, which exerts a small but noticeable effect on the small bodies on the inner edge of the disk. Some are thrown out to greater distances, but some are pulled into the inner parts of the solar system, where a few of them become visible short period comets.

While the Oort Cloud is too far away for individual objects in it to be seen from Earth, even with the largest telescopes, large objects in

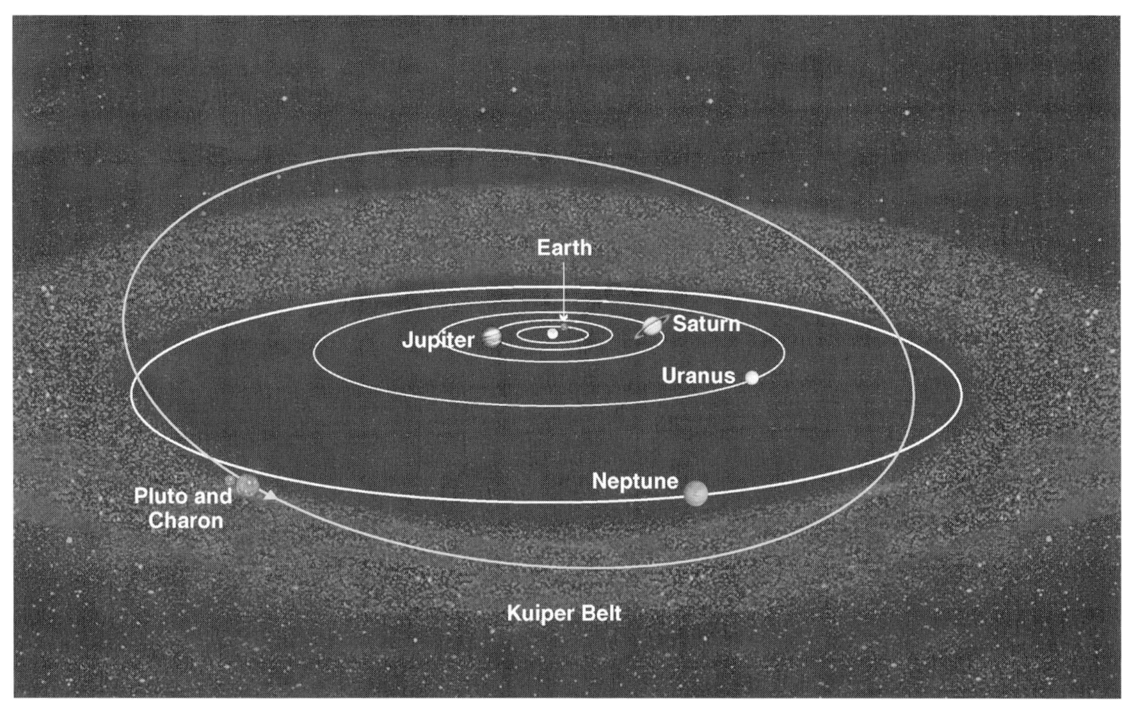

Figure 7.7 Beyond Neptune at 30 Astronomical Units from the Sun lies a recently discovered band of icy bodies from which about half of the comets originate. This donut-like distribution of icy bodies is called the Kuiper Belt, after Dutch–American astronomer Gerard P. Kuiper, who predicted its existence in 1951. The Kuiper Belt appears to extend out to about 55 AU, but some of its members have highly elongated orbits that carry them far beyond that limit. Each Kuiper Belt object is in its own orbit around the Sun, often taking as much as 500 years to make one circuit. Gravitational perturbations by Neptune, and occasional collisions among the Kuiper Belt objects can cause them to make close approaches to the Sun and inner planets, becoming an active comet. *Credit:* NASA Ames Research Center.

the much closer Kuiper Disk have been found. The first was discovered in 1992 by astronomers Jane Luu and David Jewitt, and as of mid-2004 about 900 had been identified. Most of these objects found so far are between about 60 and 600 kilometers in diameter, making them much larger than any recent comets, but smaller than the largest icy objects in the region, the planet Pluto. A few Kuiper Belt objects similar in size to Pluto's moon Charon (diameter 1250 km) have been found, while the exceptional Sedna (diameter about 1400 km) found in 2004 is nearly two-thirds the size of Pluto itself. There are an estimated 70,000 bodies larger than about 100 kilometers in diameter in the nearest parts of the Kuiper Disk, and millions more of smaller size.

It is generally agreed that objects now in the Oort Cloud were originally icy bodies that accreted from planetesimals in the region of Saturn and Uranus. The planet Jupiter formed early in the history of the solar system, and its immense gravity had a powerful effect on the formation of the other planets. The combined gravitational effects of Jupiter and the other giant planets caused a vast number of the icy and rocky planetesimals that were not pulled into the planets themselves, to be ejected to the most distant region of the newly forming solar system. The Oort Cloud was formed by this ejected material from the planetary system. The planetesimals populating the Kuiper Disk, in contrast, are thought to have formed more or less in the region where they are presently located, condensing in the outer parts of the solar nebula. It is possible, perhaps even likely, that the overall composition of the bodies in the Oort Cloud is different from the composition of the Kuiper Disk objects because of their different regions of formation in the early solar nebula.

Using the largest telescopes in the world, such as the Keck Observatory 10 meter telescopes on Mauna Kea in Hawaii, astronomers can now observe individual Kuiper Disk objects to learn something about their compositions. Astronomers are finding that Kuiper Disk objects are diverse in composition. One of them (designated 1993 SC) shows spectroscopic signs of the presence of complex hydrocarbons on its surface, probably in the form of ices, while another, 1996 TO_{66}, has the spectral signature of ordinary frozen water. Others have spectra that are essentially featureless; these objects are probably covered with a black layer of organic-rich material that hides the ices below, much as comet nuclei tend to have nearly black surfaces of similar composition.

We can also study another class of solar system object that appears to have originated in the Kuiper Disk. When Neptune pulls bodies out of the Kuiper Disk toward the inner planets, some of these objects get stuck temporarily in orbits near Saturn and Uranus. They are not usually close enough to the Sun to be warmed to the point where their ices evaporate and make them into classical comets, but instead they remain cold and frozen solid. These objects in orbits that cross the orbits of the major outer planets are called Centaurs – originally thought to be more or less ordinary asteroids, they are now generally recognized as former members of the Kuiper Disk population. The best studied Centaur is called 5145 Pholus (after one of the mythical half man, half horse Centaurs), and its surface (and probably its interior) is made of a remarkable combination of materials. First, there is frozen water and frozen methyl alcohol (CH_3OH). Also there is the mineral olivine. Perhaps most remarkable is a dark, red-colored solid material made of complex organic molecules, a mixture closely similar to a laboratory-made material called "tholin." Tholin is a term given to a group of complex organic molecular materials produced by the irradiation of a mixture of simple gases or ices by electrical discharge or

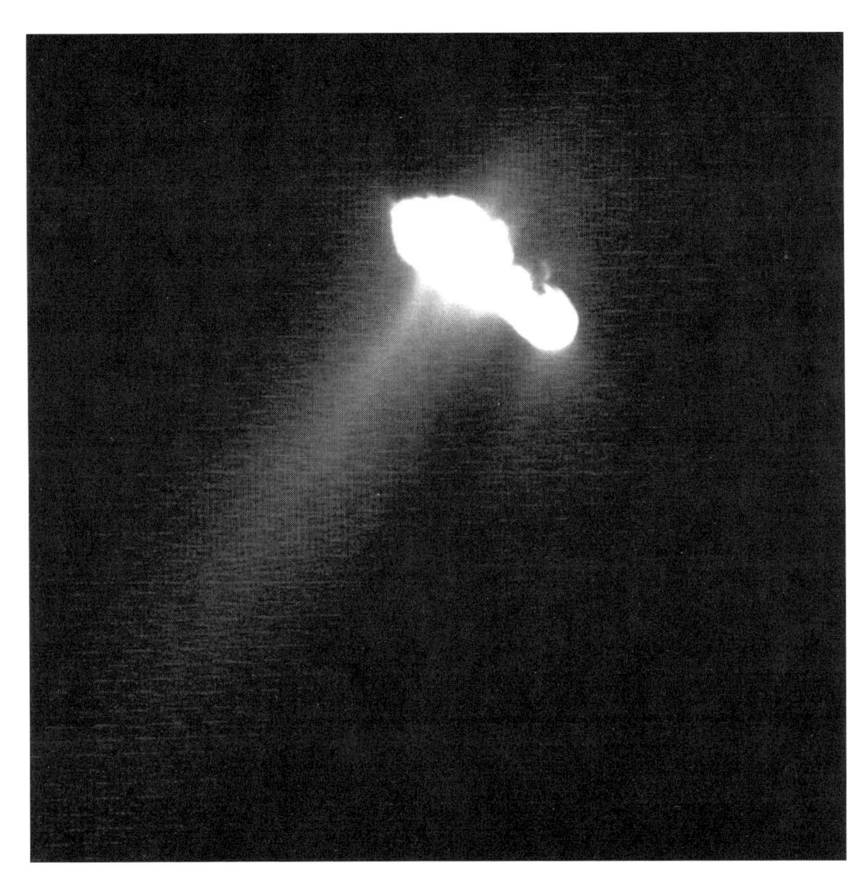

Figure 7.8 NASA's Deep Space 1 spacecraft approached Comet Borrelly on September 22, 2001, and took this image of the nucleus and a jet of dust being ejected through the porous surface of the comet. The active region on the nucleus from which this jet is emerging is about 3 kilometers long, and is on the sunlit side of the comet. The nucleus itself is about 8 kilometers long, and is overexposed in this image, which was taken to capture the dust jet. Another, smaller, jet feature is visible on the tip of the nucleus on the lower right-hand edge. The small arc of faint light at the "neck" of the nucleus appears to be a crater on the comet's surface. *Credit:* NASA/JPL/Caltech.

ultraviolet light. Tholins consist of polymers, hydrocarbons, and other combinations of the basic elements C (carbon), H (hydrogen), O (oxygen), and N (nitrogen), or CHON (as noted above), and they are thought to represent the materials produced naturally in space, as ultraviolet light interacts with ices in the interstellar clouds and the solar system.

Thus, the surface of Pholus has basically the same composition as a comet – ices, a common silicate mineral, and a complex organic solid. But what distinguishes this Centaur and the Kuiper Disk objects from traditional comets? The answer is that comets are made of the same stuff, but they come sufficiently close to the Sun for their ices to evaporate and produce the coma and tail. Most comets must come within about 5 AU (Jupiter's distance) of the Sun to warm sufficiently for evaporation to occur. At that distance a comet's surface temperature will be about 175 K (−98 C). However, some comets (such as Comet Hale-Bopp)

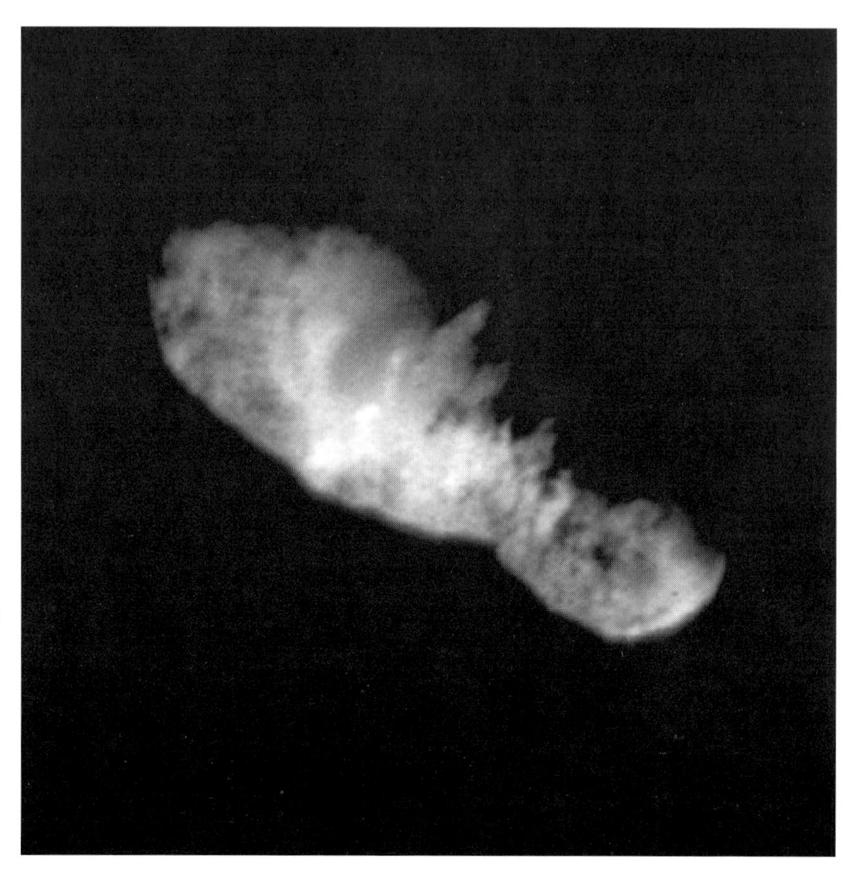

Figure 7.9 The nucleus of Comet Borelly close up. This is the highest-resolution view of the icy, rocky nucleus of this comet taken by NASA's Deep Space 1 spacecraft, just 160 seconds before the closest approach at about 3,417 kilometers. The 9 kilometer-long nucleus is dark black in color, similar to a charcoal briquette. In this enhanced image gradations of black indicate a variety of terrains and surface textures, mountains, and fault structures on the nucleus. The brightest regions are probably source regions where dust-laden gas emerges through the cracks and pores of the otherwise black crust as they rotate into sunlight as the nucleus spins.
Credit: NASA/JPL/Caltech.

begin to evaporate, or become "active" at greater distances, perhaps as much as 8 AU, where the surface temperature would be about 135 K (−138 C). The distance, or temperature at which a comet becomes active seems to depend on the kinds of ice of which it is composed (mainly H_2O and CO) and the comet's past history of passes by the Sun. Carbon monoxide is much more easily evaporated than water ice, and the activity of most comets appears to be controlled by the amount and distribution of CO ice inside them.

Space missions to comets

In addition to the Stardust mission mentioned above, several other spacecraft are being sent to comets over the next decade or more. The Deep Space 1 spacecraft passed by Comet Borelly in September 2001

(Figs. 7.8 and 7.9). The European Space Agency Rosetta spacecraft will visit Comet Churyumov/Gerasimenko in 2014–2015. The Deep Impact mission is scheduled to launch in December 2004 with the purpose of firing a projectile at the nucleus of Comet Tempel 1 to study the crater that results, and the inner layers of the comet that are exposed in the crater walls. These missions can all be followed on the web pages listed at the end of this chapter.

Conclusion

The nature of comet studies has changed markedly in the last 20 years or so. From spacecraft exploration of Comet Halley and the recent appearance of two bright comets, we have learned that comets are fundamentally made of three components: ices of many different compositions, silicate minerals, and complex organic molecules. Each component has its story to tell about the origin of the comets, the earliest days of the solar system, and even the advent of life on Earth. Through the investigation of these frigid objects we are taken back to a time before the origin of the solar system, and into the dense molecular clouds where thin layers of ice froze on microscopic grains of stardust. The insight we gain from the study of comets bears not only upon the origin of these icy transients from beyond the planetary region of the solar system, but upon our own origins as well.

Websites for comet missions and additional information about comets:

Stardust Mission: http://stardust.jpl.nasa.gov
Deep Impact Mission: http://deepimpact.jpl.nasa.gov
Rosetta Mission: http://sci.esa.int/rosetta

General information on comets, including those currently visible:
http://cometography.com/

Index

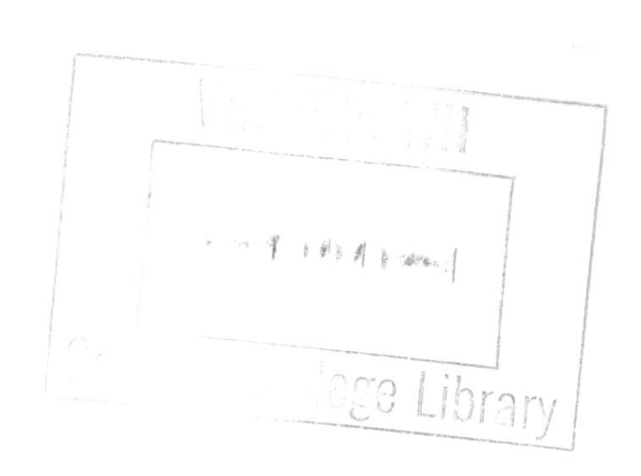
ege Library